NATIONAL ACADEMIES *Sciences*
Engineering
Medicine

PRESS
Washington, DC

Supporting Family Caregivers in STEMM

A Call to Action

Elena Fuentes-Afflick and
Katherine Wullert, *Editors*

Committee on Policies and Practices
for Supporting Caregivers Working
in Science, Engineering, and
Medicine

Committee on Women in Science,
Engineering, and Medicine

Policy and Global Affairs

Consensus Study Report

NATIONAL ACADEMIES PRESS 500 Fifth Street, NW Washington, DC 20001

This activity was supported by contracts between the National Academy of Sciences and the Doris Duke Foundation, Henry Luce Foundation, National Aeronautics and Space Administration, National Institute of Standards and Technology, National Institutes of Health (HHSN263201800029I), and U.S. National Science Foundation. Any opinions, findings, conclusions, or recommendations expressed in this publication do not necessarily reflect the views of any organization or agency that provided support for the project.

International Standard Book Number-13: 978-0-309-71358-0
International Standard Book Number-10: 0-309-71358-7
Digital Object Identifier: https://doi.org/10.17226/27416
Library of Congress Control Number: 2024938157

This publication is available from the National Academies Press, 500 Fifth Street, NW, Keck 360, Washington, DC 20001; (800) 624-6242 or (202) 334-3313; http://www.nap.edu.

Copyright 2024 by the National Academy of Sciences. National Academies of Sciences, Engineering, and Medicine and National Academies Press and the graphical logos for each are all trademarks of the National Academy of Sciences. All rights reserved.

Printed in the United States of America.

Suggested citation: National Academies of Sciences, Engineering, and Medicine. 2024. *Supporting Family Caregivers in STEMM: A Call to Action*. Washington, DC: The National Academies Press. https://doi.org/10.17226/27416.

The **National Academy of Sciences** was established in 1863 by an Act of Congress, signed by President Lincoln, as a private, nongovernmental institution to advise the nation on issues related to science and technology. Members are elected by their peers for outstanding contributions to research. Dr. Marcia McNutt is president.

The **National Academy of Engineering** was established in 1964 under the charter of the National Academy of Sciences to bring the practices of engineering to advising the nation. Members are elected by their peers for extraordinary contributions to engineering. Dr. John L. Anderson is president.

The **National Academy of Medicine** (formerly the Institute of Medicine) was established in 1970 under the charter of the National Academy of Sciences to advise the nation on medical and health issues. Members are elected by their peers for distinguished contributions to medicine and health. Dr. Victor J. Dzau is president.

The three Academies work together as the **National Academies of Sciences, Engineering, and Medicine** to provide independent, objective analysis and advice to the nation and conduct other activities to solve complex problems and inform public policy decisions. The National Academies also encourage education and research, recognize outstanding contributions to knowledge, and increase public understanding in matters of science, engineering, and medicine.

Learn more about the National Academies of Sciences, Engineering, and Medicine at **www.nationalacademies.org**.

Consensus Study Reports published by the National Academies of Sciences, Engineering, and Medicine document the evidence-based consensus on the study's statement of task by an authoring committee of experts. Reports typically include findings, conclusions, and recommendations based on information gathered by the committee and the committee's deliberations. Each report has been subjected to a rigorous and independent peer-review process and it represents the position of the National Academies on the statement of task.

Proceedings published by the National Academies of Sciences, Engineering, and Medicine chronicle the presentations and discussions at a workshop, symposium, or other event convened by the National Academies. The statements and opinions contained in proceedings are those of the participants and are not endorsed by other participants, the planning committee, or the National Academies.

Rapid Expert Consultations published by the National Academies of Sciences, Engineering, and Medicine are authored by subject-matter experts on narrowly focused topics that can be supported by a body of evidence. The discussions contained in rapid expert consultations are considered those of the authors and do not contain policy recommendations. Rapid expert consultations are reviewed by the institution before release.

For information about other products and activities of the National Academies, please visit www.nationalacademies.org/about/whatwedo.

COMMITTEE ON POLICIES AND PRACTICES FOR SUPPORTING CAREGIVERS WORKING IN SCIENCE, ENGINEERING, AND MEDICINE

ELENA FUENTES-AFFLICK, M.D., M.P.H. (*Chair*) (NAM),[1] Professor of Pediatrics, University of California, San Francisco; Vice Dean, UCSF School of Medicine at Zuckerberg San Francisco General Hospital and Trauma Center

MARIANNE BERTRAND, Ph.D. (NAS), Chris P. Dialynas Distinguished Service Professor of Economics, University of Chicago; Research Fellow, National Bureau of Economic Research, Center for Economic Policy Research, and Institute for the Study of Labor

MARY BLAIR-LOY, Ph.D., Professor of Sociology, University of California, San Diego; Co-Director, Center for Research on Gender in STEMM

KATHLEEN CHRISTENSEN, Ph.D., Faculty Fellow, Boston College Center for Social Innovation; Founder/Director, Alfred P. Sloan Foundation Program on Workplace, Workforce and Working Families Program

J. NICHOLAS DIONNE-ODOM, Ph.D., R.N., Associate Professor of Nursing, University of Alabama at Birmingham; Co-Director, Bereavement Support Services, UAB Center for Palliative and Supportive Care

MIGNON DUFFY, Ph.D., Professor of Sociology, University of Massachusetts Lowell

JEFFREY GILLIS-DAVIS, Ph.D., Professor of Physics, Washington University in St. Louis

RESHMA JAGSI, M.D., D.Phil., Chair, Department of Radiation Oncology, Emory University and Winship Cancer Institute

ELLEN ERNST KOSSEK, Ph.D., Basil S. Turner Distinguished Professor, Mitchell E. Daniels Jr. School of Business, Purdue University; Former President, Work-Family Researchers Network

LINDSEY MALCOM-PIQUEUX, Ph.D., Assistant Vice President for Diversity, Equity, Inclusion, and Assessment and Chief Institutional Research Officer, California Institute of Technology

---

[1] Designates membership in the National Academy of Sciences (NAS), National Academy of Engineering (NAE), or National Academy of Medicine (NAM).

SANDRA KAZAHN MASUR, Ph.D., Professor of Ophthalmology and of Pharmacological Sciences and Director, Office for Women's Careers, Icahn School of Medicine at Mount Sinai
MARIA ONG, Ph.D., Senior Research Scientist, TERC
ROBERT L. PHILLIPS, JR., M.D., M.S.P.H. (NAM), Founding Executive Director, The Center for Professionalism and Value in Health Care
JASON RESENDEZ, President and CEO, National Alliance for Caregiving
HANNAH VALANTINE, M.D., M.B.B.S. (NAM), Professor of Medicine, Stanford University; Inaugural Chief Officer for Scientific Workforce Diversity, National Institutes of Health
JOAN WILLIAMS, J.D., Sullivan Professor of Law and Founding Director, Center for WorkLife Law, University of California Hastings College of the Law, San Francisco

Staff

KATHERINE WULLERT, Ph.D., Study Director and Program Officer, Committee on Women in Science, Engineering, and Medicine
ASHLEY BEAR, Ph.D., Director, Committee on Women in Science, Engineering, and Medicine
ABIGAIL HARLESS, Research Associate, Committee on Women in Science, Engineering, and Medicine
PAMELA LAVA, Senior Program Assistant, Committee on Women in Science, Engineering, and Medicine

Consultants

NGOC DAO, Ph.D., Associate Professor, College of Business and Public Management, Kean University
ERIN FRAWLEY, M.Ed., Education Equity Program Manager, Center for WorkLife Law
JESSICA LEE, J.D., Senior Staff Attorney, Center for WorkLife Law; Director, Pregnant Scholar Initiative
ASHLEY LOWE, M.P.H., Researcher, Transformative Research Unit for Equity, RTI International

JENNIFER LUNDQUIST, Ph.D., Associate Dean of Research and Faculty Development and Professor of Sociology, University of Massachusetts Amherst
TASSELI McKAY, Ph.D., Researcher, Transformative Research Unit for Equity, RTI International
JOYA MISRA, Ph.D., Provost Professor and Roy J. Zuckerberg Endowed Leadership Chair, Department of Sociology, University of Massachusetts Amherst
JOANNA RICCITELLI, Ph.D. Candidate in Sociology, University of Massachusetts Amherst
MONICA SHEPPARD, M.S.W., Co-Director, Emerging Equities Scholar Program, Transformative Research Unit for Equity, RTI International
SARAH STOLLER, Ph.D., Freelance Writer, Editor, and Research Consultant
COURTNEY VAN HOUTVEN, Ph.D., Professor, Department of Population Health Science, Duke University

COMMITTEE ON WOMEN IN SCIENCE, ENGINEERING, AND MEDICINE

GILDA BARABINO, Ph.D. (*Chair*) (NAE, NAM), President and
 Professor of Biomedical and Chemical Engineering, Olin College
SANDRA BEGAY, M.S., Principal Member of the Technical Staff, Sandia
 National Laboratories
JOAN WENNSTROM BENNET, Ph.D., Distinguished Professor of
 Plant Biology and Pathology, Rutgers University
MAY R. BERENBAUM, Ph.D., Professor and Head of Entomology
 and Professor of Plant Biology, University of Illinois at Urbana-
 Champaign
VALERIE CONN, Founder and Principal, Future Science Now
LESLIE D. GONZALES, Ph.D., Associate Professor, Higher, Adult,
 and Lifelong Learning Unit, College of Education, Michigan State
 University
EVELYNN M. HAMMONDS, Ph.D., Barbara Guttman Rosenkrantz
 Professor of the History of Science and of African and African
 American Studies, Harvard University
RESHMA JAGSI, M.D., D.Phil., Chair, Department of Radiation
 Oncology, Emory University and Winship Cancer Institute
HILLARY LAPPIN-SCOTT, Ph.D., Honorary Distinguished Professor
 of Microbiology, Cardiff University
MANUEL PÉREZ-QUIÑONES, Ph.D., Professor of Software and
 Information Systems, University of North Carolina at Charlotte
REED V. TUCKSON, M.D., F.A.C.P., Managing Director, Tuckson
 Health Connections, LLC
ELENA FUENTES-AFFLICK, M.D., M.P.H. (*Ex-officio member*)
 (NAM), Professor of Pediatrics, University of California, San
 Francisco; Vice Dean, UCSF School of Medicine at Zuckerberg San
 Francisco General Hospital and Trauma Center
CAROL K. HALL, Ph.D. (*Ex-officio member*) (NAE), Distinguished
 University Professor of Chemical and Biomedical Engineering, North
 Carolina State University
SUSAN R. WESSLER, Ph.D. (*Ex-officio member*) (NAS), Distinguished
 Professor of Genetics and Neil and Rochelle Campbell Chair for
 Innovation in Science Education, University of California, Riverside

Reviewers

This Consensus Study Report was reviewed in draft form by individuals chosen for their diverse perspectives and technical expertise. The purpose of this independent review is to provide candid and critical comments that will assist the National Academies of Sciences, Engineering, and Medicine in making each published report as sound as possible and to ensure that it meets the institutional standards for quality, objectivity, evidence, and responsiveness to the study charge. The review comments and draft manuscript remain confidential to protect the integrity of the deliberative process.

We thank the following individuals for their review of this report:

JESSICA CALARCO, University of Wisconsin–Madison
PAULA ENGLAND, New York University, Abu Dhabi
NANCY FOLBRE, Bard College
RASHEED GBADEGESIN, Duke University
WILLIAM HALEY, University of South Florida
HOPE ISHII, University of Hawai'i at Mānoa
ARLENE KATZ, Harvard University
NWANDO OLAYIWOLA, Humana
CAROL THOMSEN, All Five
ISABEL TORRES, Mothers in Science
JIANYU XU, Purdue University

Although the reviewers listed above provided many constructive comments and suggestions, they were not asked to endorse the conclusions or recommendations of this report nor did they see the final draft before its release. The review of this report was overseen by JOAN BENNETT, Rutgers University–New Brunswick, and BETTY FERRELL, City of Hope National Medical Center. They were responsible for making certain that an independent examination of this report was carried out in accordance with the standards of the National Academies and that all review comments were carefully considered. Responsibility for the final content rests entirely with the authoring committee and the National Academies.

Preface

Most of us have firsthand experience with caregiving, yet family caregiving remains a taboo topic in many sectors of academic science, technology, engineering, mathematics, and medicine (STEMM). Faculty, researchers, staff, students, and trainees manage a range of caregiving responsibilities, yet caregiving is rarely discussed in the workplace and the policies to support family caregivers are often limited.

Beginning in January 2023, the Committee on Policies and Practices for Supporting Family Caregivers Working in Science, Engineering, and Medicine of the National Academies of Sciences, Engineering, and Medicine convened to address the issues experienced by caregivers in academic STEMM. Throughout the process of preparing this report, the universality and challenges of caregiving were abundantly clear. Caregiving responsibilities arose at multiple instances during the process, whether committee members or staff members needed to care for sick parents, provide care for children or grandchildren, or balance childcare and everyday work. Speakers shared their caregiving experiences related to sick relatives and caregiving at a distance during our public symposia held in February and March of 2023. These experiences underscored the need for the committee's evaluation and highlighted the challenge of making effective, actionable recommendations.

The committee's work was grounded in the goals of the National Academies' Committee on Women in Science, Engineering, and Medicine (CWSEM), which championed this project. As framed by CWSEM, there is no single prototype for the role of family caregiver, but women, particularly women of color, are disproportionately affected by caregiving responsibilities. Thus, advancing gender and racial equity in STEMM requires addressing the needs and challenges of family caregiving. It is our hope that addressing the needs of family caregivers will promote equity and ensure a more flexible, inclusive, and welcoming environment for everyone in the academic STEMM workforce.

This report is the culmination of 18 months of work from an engaged, diverse committee of experts who demonstrated a deep commitment to family caregiving. Our work would not have been possible without the support of the National Institutes of Health, the National Science Foundation, the National Aeronautics and Space Administration, the National Institute of Standards and Technology, the Doris Duke Charitable Foundation, and the Henry Luce Foundation. These sponsors provided resources to the committee as well as their time and expertise throughout the committee's deliberations.

The committee hopes that the report stimulates discussion and ignites more care about caregiving through action at societal, governmental, and institutional levels.

I hope that this report will motivate leaders, colleges, and universities to foster and facilitate greater discussion of caregiving in academic STEMM, ensure that their policies and practices support family caregivers, and accelerate progress toward an optimal STEMM workplace. Ultimately, we seek broad, lasting culture change to support family caregivers.

Elena Fuentes-Afflick, *Chair*
Committee on Policies and Practices for Supporting Family Caregivers Working in Science, Engineering, and Medicine

Contents

	SUMMARY	1
1	COMMITTEE TASK AND APPROACH	13
	Activities to Inform the Report, 18	
	Organization of the Report, 20	
2	OVERVIEW OF UNPAID FAMILY CAREGIVING	23
	What Is Family Caregiving? A Typology of Care, 23	
	Who Are Family Caregivers?, 27	
	Rising National Trends in Family Caregiving Needs, 30	
	The Impact of Caregiving and the Challenges Faced by Caregivers, 34	
	Summary of Findings from Chapter 2, 40	
3	CAREGIVING CHALLENGES AND IMPLICATIONS FOR EQUITY IN STEMM	43
	Ideal Worker Norm and Work Devotion Schema in STEMM, 44	
	Consequences of Ideal Worker and Work Devotion Norms to STEMM Innovation and Vitality, 46	
	Bias and Discrimination Against Caregivers, 48	
	The Impact of Cultural Schemas on the Organization of STEMM, 55	

Impacts of COVID-19: Exacerbating Longstanding
  Inequities in the Provision of Care, 57
Balancing the Benefits and Drawbacks of Ideal Worker Norms, 59
Summary of Findings from Chapter 3, 60

4 CURRENT LAWS, POLICIES, AND PRACTICES TO
  SUPPORT FAMILY CAREGIVERS 63
  Current Federal and State Laws, 63
  Federal Agencies' and Other Funders' Policies
    Supporting Caregivers, 67
  Accreditation and Certification Boards, 69
  Institutional Policies, 70
  Summary of Findings from Chapter 4, 78

5 BARRIERS TO EFFECTIVE POLICY
  IMPLEMENTATION 81
  Affordability, 81
  Availability, 83
  Lack of Awareness, 84
  Lack of Attention to Intersectionality, 86
  Lack of Institutionalization, 88
  Cultural Beliefs and Biases, 90
  Summary of Findings from Chapter 5, 92

6 BEST PRACTICES FOR COLLEGES AND
  UNIVERSITIES 95
  Foundational Minimums for Legal Compliance, 96
  Best Practices for Institutional Policies, 102
  Cultural Shifts to Challenge Ideal Worker Norms, 117
  Summary of Findings from Chapter 6, 119

7 INNOVATIVE APPROACHES TO CAREER
  FLEXIBILITY 121
  Innovative Approaches to Promote Workplace Flexibility, 125
  Summary of Findings from Chapter 7, 134

8 RECOMMENDATIONS AND CONCLUSIONS 135
  Recommendations for Colleges and Universities, 135
  Recommendations for Federal and Private Funders, 142
  Recommendations to Congress and the Federal Government, 143

REFERENCES 145

APPENDIXES
A Biographical Sketches of Committee Members and
   Commissioned Paper Authors 171
B Caregiver Interview Sample and Methodology 185
C Methodology for Selecting Causal Analyses of the Economic
   Impact of Caregiving 193
D Current Federal, State, and Local Policies to Support
   Family Caregivers 195

Boxes, Figures, and Tables

BOXES

1-1 Statement of Task, 14
1-2 Titles and Positions Within Academic STEMM, 17

6-1 Checklist for Training to Create an Environment Where University Staff and Leaders Understand Caregivers' Legal Rights, 98
6-2 Examples in Action: Make Policies and Resources Easy to Find and Access, 101
6-3 Best Practices Related to Caregiving Leave Checklist, 103
6-4 Examples in Action: Retaining Clinical Scientists, 105
6-5 Best Practices for Policies Related to Accommodations and Adjustments Checklist, 105
6-6 Examples in Action: Supporting Postdoctoral Caregivers, 106
6-7 Best Practices for Policies Related to Direct Care Support Checklist, 110
6-8 Example in Action: Backup Care, 110
6-9 Example in Action: Supporting Pregnant and Parenting Students, 111
6-10 Best Practices for Caring for Adult Dependents Checklist, 114
6-11 Example in Action: Supporting Graduate Student and Trainee Caregivers, 115

6-12 Example in Action: Adult Care, 116

7-1   Reimagined Academy: Alternative Visions of
      Academic Success, 122
7-2   Reimagined Academy: Care-Centered Academic
      Workplace Norms, 124

FIGURES

2-1   A typology of care, 25

B-1   Career stages, 191

TABLES

2-1   Causal Effects of Caregiving on Key Economic Outcomes, 36

4-1   Federal and State Legal Protections for Employees, 64
4-2   Federal and State Legal Protections for Students, 66

B-1   Family Codes, 188
B-2   Deductive Codes, 190

C-1   Work Outcomes, 194

Summary

Care is one of the most universal human experiences. We all need care as babies and young children, at times in our lives when we are ill, or as we age. Most of us provide family care, largely unpaid, throughout our lives, whether this takes the form of parenting, caring for older adults, or caring for relatives or loved ones who are ill or have a disability. Caregiving is necessary and important labor that helps society to function and thrive, and proper support for caregivers is crucial for national economic growth, economic and social outcomes for families, and gender equality. However, stigma and barriers exist within the U.S. workforce for family caregivers,[1] including those pursuing a career in science, technology, engineering, mathematics, and medicine (STEMM). These stigma and barriers disproportionately affect women and contribute to the national trend of fewer women advancing and succeeding in academic STEMM careers. The U.S. STEMM workforce faces a lack of governmental and organizational support

---

[1] The term *family caregivers* is used throughout this report to refer to those who provide largely unpaid care to family members, friends, and loved ones. Other terms used in the literature included *unpaid caregivers* or *informal caregivers*. This definition includes those caring for children as well as for adults with illnesses or disabilities. The committee acknowledges that the current literature on caregiving varies in the types of labor that are included and thus can produce disparate estimates of both the magnitude and the intensity of care. The committee adopted a broad definition to capture the full expanse of what caregiving includes (see Chapter 2), but also provides details on how caregiving is defined when providing estimates of the caregiving population and time spent on caregiving.

for caregiving that needs to be addressed to ensure inclusion and continued innovation and competitiveness.

This report aims to capture the ways in which the labor and contributions of caregivers are often invisible and undervalued, with a specific focus on the academic STEMM ecosystem, including undergraduate and graduate students, postdoctoral scholars, resident physicians and other trainees, tenure-track and non-tenure-track faculty, staff, and researchers. The report describes how caregiving responsibilities clash with ingrained norms in academic STEMM environments, which demand that STEMM students and workers demonstrate immense devotion to their fields and are always available and visibly working. The report reviews policies and practices that support caregivers, locally and nationally, and describes best practices in policy implementation and design. It also highlights innovative practices and offers actionable recommendations to higher education institutions, public and private funders, and the federal government.

The goal of this report and its recommendations is to facilitate and accelerate greater participation of caregivers in STEMM education and work and thereby advance scientific innovation and support a stronger and more inclusive academic STEMM workforce. The academic STEMM workforce needs caregivers, and, particularly at a time when many STEMM fields face challenges with workforce shortages and a lack of diversity, support for this diverse population is even more important. Family caregiving is not simply an outside obligation that has no bearing on the workings of academic STEMM as it affects the lives of so many people working and studying in colleges and universities around the country. The committee seeks with this report to showcase the immense value caregivers bring to academic STEMM, the current limitations of the system to adequately support them, and the kinds of solutions that will create a more welcoming and inclusive STEMM environment.

CAREGIVING AND STEMM

Family caregivers make up an important and significant portion of the U.S. workforce and the STEMM workforce. Though in the past, most caregivers were women who stayed at home, today that has shifted and millions of people who provide care are also employed either part-time or full-time (Lerner, 2022). Estimates from 2019 find that, among the nearly 53 million adults in the United States providing care for someone aged 18 or older, more than 60 percent were employed (AARP & National Alliance for Caregiving, 2020). Further, examinations of workers with graduate or postgraduate degrees find that around 15 percent of these workers are caregivers (Cynkar & Mendes, 2011). Caregivers are the faculty, researchers, postdocs, students, and staff that make up academic STEMM. As such, a lack of support for caregiving has consequences not just for individual caregivers but for the STEMM enterprise as a whole.

Research has shown the significant effect of a lack of support for family caregiving on the careers of STEMM professionals. For example, a 2019 study using nationally representative data on full-time professionals in science, technology, engineering, and mathematics found that a startling 43 percent of new mothers and 23 percent of new fathers left full-time employment in these fields after the birth of their first child, and most of the mothers and a substantial portion of fathers noted they left for "family-related" reasons (Cech & Blair-Loy, 2019). Other work has found that scientists may feel the need to have fewer children than they would otherwise desire given the demands of STEMM, and having fewer children than desired is related to lower satisfaction and greater plans to exit STEMM (Ecklund & Lincoln, 2011). The impacts of a lack of caregiving support on attrition are problematic especially in fields already facing challenges with diversity and hiring shortages (American Association of Colleges of Nursing, 2023; Jean, Payne, & Thompson, 2014).

LACK OF SUPPORT AND THE IMPACT ON CAREGIVERS

Caregiving of all forms can be rewarding and individually fulfilling for the caregiver and care recipient, producing strengthened ties with loved ones, a sense of purpose, and other positive emotions (AP-NORC Center for Public Affairs Research, 2014; Hoefman et al., 2013; Mackenzie & Greenwood, 2012; Quinn et al., 2010). However, there are also economic and physical challenges and emotional burdens associated with caregiving and the lack of institutional and societal support for caregivers. Individual caregivers can incur heavy financial costs and reduced earnings and advancement at work (Wakabayashi & Donato, 2005). Family caregivers who take reduced hours or leave the workforce incur foregone wages and reduced retirement savings (Weller & Tolson, 2019) as well as the potential for career disruptions and loss of seniority (Bainbridge & Broady, 2017; Stoner & Stoner, 2016). Those who do not or cannot leave the workforce and lack sufficient support at work face potential constraints on productivity (Morgan et al., 2021), disruptions to work schedules (AARP & National

Alliance for Caregiving, 2020; Witters, 2011), and discrimination, which has increased over the past few decades (Morris et al., 2021). Family caregiving can also take a physical and emotional toll on the individual providing care, with caregivers reporting greater levels of anxiety and depression; insufficient sleep; worse self-reported health; and limited time for exercise, rest, and personal care of their own physical and mental health needs (Centers for Disease Control and Prevention, 2019; Conley et al., 2004; Park, 2020; Tay et al., 2022).

The barriers that family caregivers face in academic STEMM fields are further exacerbated by expectations of devotional allegiance to work (Blair-Loy & Cech, 2022). Many academic STEMM environments uphold "ideal worker" norms, norms that suggest that to be an ideal worker[2] requires full dedication to work such that a person's life centers on their work without outside influences affecting them (Acker, 1990; Williams, 1989). In these settings, strong stigma develops around any apparent violations of these norms. In particular, flexibility stigma[3] is one such censure of those who go against ideal worker norms by utilizing or appearing to utilize accommodations and alternative work arrangements that allow them to attend to responsibilities outside of work (Williams et al., 2000). The entrenched nature of ideal worker norms also posed additional challenges during the COVID-19 pandemic, as many caregivers, particularly of young children, struggled to navigate the high expectations and demands of paid work alongside increased caregiving demands at home while working (Zanhour & Sumpter, 2022).

CAREGIVING AND EQUITY

Family caregiving cuts across gender, race/ethnicity, and other characteristics, and a lack of support for caregiving poses challenges for employees with caregiving responsibilities from all backgrounds. At the same time, women continue to bear disproportionate caregiving responsibilities in

---

[2] Ideal worker norms grew from the separation of work and home into distinct spheres coupled with the development of new technologies through which employers could track worker productivity with the onset of the industrial revolution. These norms were adapted in white-collar work in the early 20th century as these settings took on productivity standards seen prior in factories. Media portrayals in the following decades further entrenched these norms (Davies & Frink, 2014).

[3] *Flexibility stigma* refers to negative evaluations and/or treatment of individuals who make use of policies designed to allow greater flexibility in work schedule, location, or intensity.

the United States (U.S. Bureau of Labor Statistics, 2023a). This societal distribution of care has substantial implications for gender equity generally and is of growing concern in academic STEMM roles. The burden of these family caregiving responsibilities placed on women contributes to unequal advancement opportunities and influences career choices (Fox & Gaughan, 2021; Wakabayashi & Donato, 2005). Additionally, research has shown that even for women who may not have caregiving responsibility, the "specter of motherhood," or the belief that all women want to and will become mothers, looms and leads to presumptions about their long-term engagement and commitment to STEMM (Thébaud & Taylor, 2021).

The challenges faced by mothers and women caregivers may also be particularly acute for women of color given intersecting biases of gender and race/ethnicity (Kachchaf et al., 2015; Williams, 2014). Additionally, policies to support caregivers are also shaped by assumptions that more often align with the experiences of White, middle- to upper-class Americans but less so with other groups. For example, while many existing caregiving policies assume the care provider is a direct relative of the care recipient, Black, Hispanic, and Asian caregivers are more likely to take care of nonrelatives or extended family members (McCann et al., 2000; Sodders et al., 2020).

CURRENT CAREGIVING SUPPORTS AND BARRIERS

Supports for caregivers in academic STEMM are generally piecemeal and incomplete, creating challenges for family caregivers. A patchwork of federal and state legal requirements exists by which universities and other organizations must abide. There is no federal law establishing a right to paid leave, and because there are so many laws operating, the legal landscape is quite complex, making it challenging to navigate and understand available protections. The complexity may also contribute to the high degree of noncompliance seen across universities (Calvert, 2016; Gulati et al., 2022; Lee et al., 2017; Mensah et al., 2022; Williams et al., 2022). Current laws are also incomplete, often focusing on caregivers of young children and not as frequently considering protections and supports for caregivers of adults.

Caregivers in STEMM are also influenced by the policies of funders and accrediting institutions. Federal agencies and private funders, for example, may provide flexibility in timelines as well as use of grant monies to support family caregiving responsibilities. Federal agencies and private funders may also provide guidance to the institutions they fund to better support the STEMM workforce. Accrediting bodies also set standards for university family caregiving support.

Finally, colleges and universities have a myriad of local policies. Common policies include caregiving leave, accommodations and adjustments, direct care support, and protections against discrimination. Though support for family caregiving can exist in these many forms, there is still limited knowledge on exactly how many institutions provide each of these policies and programs and immense variation in how much support is offered. Ultimately, many caregivers in academic STEMM find current support lacking to meet their needs. This report aims to reduce this variation that leads some family caregivers to encounter less support than others and to outline policies and practices that are greatly needed to produce a more inclusive academic STEMM ecosystem. It calls for concerted action to support family caregivers so that they can have fulfilling careers and thrive in academic STEMM.

CONCLUSIONS

To this end, the committee reached five major conclusions based on a comprehensive review of the literature detailed across the chapters that follow.

1. **Supporting family caregivers is an issue of equity and a strategic labor force investment.**
   Because family caregiving demands are unevenly distributed, women have borne the brunt of the financial, mental, and physical burdens of caregiving. Though current data on caregiving and race/ethnicity as well as the intersection of race/ethnicity and gender is much more limited, existing evidence suggests that women of color may face a particularly strong burden. Lack of support for family caregivers and stigma against caregiving can push women further out of academic STEMM, an environment where women are already underrepresented. The loss of a STEMM academic due to insufficient caregiving support results in turnover, inefficiencies due to hiring replacements, increased organizational stress due to understaffing and workforce churn, the loss of training, experience, and hundreds of thousands of dollars, often from federally funded grants. It also perpetuates less STEMM workforce diversity, and it

risks ongoing labor shortages, impoverishing growth, innovation, creativity, solutions, and success.

2. **Family caregivers provide care in many forms and for a wide range of relationships, but family caregiving is often viewed in a very limited way.**

   Family caregiving includes individuals of all ages with a variety of relationships to their care provider—children, adult dependent children, parents, spouses, nonrelative loved ones, and neighbors. It entails a wide range of tasks, including caring for physical or mental health needs, providing transportation and organizational support, and assisting with finances. Family caregiving also can vary both across individuals and over time as to whether the tasks are ongoing or short-term and whether they involve more or less intense effort from the caregiver. This range is not often considered in conversations about caregiving or the policies and programs that are offered. Most often, family caregiving is implicitly or explicitly viewed as largely parents caring for children or children caring for aging parents, which can limit the sufficiency of policies. A broad understanding of caregiving is a key component to ensuring policies fully support the wide variety of needs of family caregivers.

3. **Cultural barriers present a particular challenge for increasing support for family caregivers in academic STEMM.**

   The culture of academic STEMM sets expectations for ideal workers who are expected to be able to devote immense time, energy, and attention to work without being affected by outside demands. This presents challenges for family caregivers and results in flexibility stigma, a form of bias and discrimination that penalizes those who need to seek out resources and supports that allow them to meet needs outside of paid work. These cultural stigmas affect people of all backgrounds but can be particularly detrimental to women given both the disproportionate family caregiving burden they bear as well as assumptions made about caregiving and gender. Efforts to challenge these assumptions are needed to shift cultural norms and set a more level playing field.

4. **Policy options and best practices exist for colleges and universities to support caregivers, but many institutions still fail to meet their needs.**

   Exemplar policies and practices have been implemented at colleges and universities to support caregivers covering many different needs, including leave, accommodations and adjustments, direct care support such as on-site centers, and protection from discrimination and bias. These are not consistently employed across academic institutions, and issues of availability and affordability produce barriers. When implemented, existing policies are often not well communicated, leaving people unaware of what they can access. Additionally, many policies are written without attention to the wide variety of caregiving experiences as well as differences across caregivers along lines of race/ethnicity, gender, and other characteristics that influence what caregiving looks like. Moreover, even when direct care supports are provided such as on- or near-site child or older adult care, they are often underresourced with limited availability and long waiting lists. Even with these challenges, best practices can and should be implemented with attention to evaluation to understand their effect and potential unintended consequences.

5. **Federal and state regulations cover many sets of needs, but they remain incomplete and piecemeal and are not always followed.** Unlike other industrialized nations, the United States still does not provide federal protections for paid family and sick leave. It is in fact the only country in the Organisation for Economic Co-operation and Development that does not have a national, paid caregiving leave policy. Some federal and state laws governing caregiving already exist and more are emerging across the country, encompassing the right to caregiving leave, protections against discrimination, and access to other supports such as maternity care and lactation support. The legal landscape, however, is composed of a disconnected set of mandates that is quite complex to follow and understand. This in part may contribute to why institutions often fail to meet their legal obligations to caregivers. All parts of the STEMM workforce are affected: students and faculty, as well as postdocs, medical residents and interns, and staff in soft-money research positions, emphasizing the need for centralized government action.

6. **Innovation in caregiving support is desperately needed.** Given continued challenges and gaps as well as persistent barriers to policies, innovation is needed to support family caregivers. The policies supporting caregiving that already exist do not sufficiently address common situations that undermine the vitality

and effectiveness of the STEMM workforce. These challenges have been underappreciated and include the needs of those caring for adults or nonrelative loved ones. There are innovative solutions that need greater implementation and evaluation to see how they could strengthen, augment, and expand upon existing support. Unfortunately, senior leaders in STEMM, policymakers, and the public still seem unaware of the urgent need or uncertain how to prioritize organizational innovation and flexibility to enhance support of family caregivers as a critical issue for the future of the STEMM workforce and the nation's capacity to remain a STEMM leader in the world. This cannot, however, stand in the way of creative and new ideas to produce more effective policies, practices, and interventions to support family caregivers.

RECOMMENDATIONS

The committee's recommendations focus on tangible actions that need to be taken by universities, federal and private funders, and Congress and the federal government to ensure adequate support for family caregivers in academic STEMM. These recommendations address needs not only for leave, flexibility, and direct care support, but also for a greater understanding of the efficacy of current efforts and support for innovations to better assist family caregivers. Greater detail and specific guidance on implementing these recommendations can be found in Chapter 8.

University Recommendations

Universities must and can do a great deal to support family caregivers. The overarching goal of these recommendations is to help universities create an environment that allows for continued and sustainable productivity in a way that is more inclusive of family caregivers. Such an environment shows a continued commitment to the long-term health and well-being of the academic STEMM workforce and challenges ideals of overwork as well as barriers to needed leave and flexibility. This overarching goal is reflected throughout these recommendations, which provide individual, concrete steps that can be taken. Together, they can serve to shift broader cultural norms in more inclusive ways. To assist in goal setting to achieve a more caregiver-friendly workplace, the committee organized its recommendations to universities into three categories: legal compliance, best practices, and innovative practices.

Legal Compliance

RECOMMENDATION 1: To ensure accountability and compliance, college and university leadership need to appoint a senior leader, ombuds, or team who is responsible for protecting, publicizing, and monitoring compliance with the legal mandates under Title IX, Title VII, the Family Medical and Leave Act, the Pregnant Workers Fairness Act, and any state- and local-level policies that protect caregiving faculty, postdocs and other trainees, students, and staff.

Best Practices

RECOMMENDATION 2: *Caregiving Leave.* Colleges and universities should comply with FMLA's requirement for 12 weeks of unpaid leave per year and provide paid family and medical leave to faculty, staff, postdocs and other trainees, and graduate students receiving pay, even if this leave is not mandated by state or federal law. Additionally, colleges and universities should provide leave for caregiving students, which allows them to maintain their student status so that they can continue to receive any aid or health insurance to which they are entitled.

RECOMMENDATION 3: *Accommodations and adjustments.* Colleges and universities should institutionalize opportunities for individually customized work and educational flexibility across a variety of needs, including location, time, workload, and intensity.

RECOMMENDATION 4: *Direct care support.* Centralized resources to support basic caregiving needs for staff, faculty, postdocs and other trainees, and students need to be easily available and searchable.

RECOMMENDATION 5: *Data Collection and Analysis.* To ensure that colleges and universities understand the needs of the caregiving populations within their ranks, understand the impact of their policies, existing and new, and address potential unintended consequences, colleges and universities should collect and analyze data on family caregivers.

Innovative Practices

RECOMMENDATION 6: Colleges and universities should pilot and evaluate innovative policies and practices intended to increase support for caregivers and influence lasting cultural change. Less research-intensive colleges and universities should partner with research-intensive institutions and participate in projects and efforts to test new policy ideas.

Federal Agencies and Private Funders

RECOMMENDATION 7: Federal and private funders should allow and support flexibility in the timing of grant eligibility as well as grant application and delivery deadlines for those with caregiving responsibilities and provide support for coverage while a grantee is on caregiving leave.

RECOMMENDATION 8: Federal and private funders should facilitate the leave and reentry processes for those who take a caregiving leave.

RECOMMENDATION 9: Federal and private funders should fund innovative research on family caregiving in academic STEMM by providing competitive grants to institutions to support pilot projects and develop policy innovations. Funders should collaboratively develop and offer caregiver policy guidance to the institutions they fund based on the findings of this research as well as existing evidence.

Congress and the Federal Government

RECOMMENDATION 10: Congress should enact legislation to mandate a minimum of 12 weeks of paid, comprehensive caregiving leave. This leave should cover the various forms of caregiving, including childcare, older adult care, spousal care, dependent adult care, extended family care, end-of-life care, and bereavement care.

RECOMMENDATION 11: Following the model of the recent CHIPS and Science Act, which required the provision of on-site childcare for those seeking access to funds supporting semiconductor development, the agency or department tasked with implementation of future STEMM funding bills should include support for childcare in application requirements.

1

Committee Task and Approach

In December 2019, the National Academies of Sciences, Engineering, and Medicine's standing Committee on Women in Science, Engineering, and Medicine held a public workshop to scope the current state of knowledge on the impact of caregiving responsibilities on gender equity in science, engineering, and medicine and the policy landscape available to support caregivers. The public workshop revealed a range of serious issues—widespread bias and discrimination against caregivers; a lack of comprehensive policies and resources at local, national, and organizational levels; poor implementation; and underutilization of existing policies at institutions. At that time, the committee could not foresee the global pandemic that would come a year later and further exacerbate the longstanding challenges facing family caregivers in science, technology, engineering, mathematics, and medicine (STEMM) fields. The many challenges facing caregivers that were made more visible during the COVID-19 pandemic, coupled with the information shared at the workshop, reinforced for the committee the critical need for a major National Academies consensus study to detail the current challenges caregivers face and provide clear recommendations to better support family caregivers in academic STEMM.

Thus, in September 2022, with support from a coalition of private and public sponsors, including the National Institutes of Health, the National Science Foundation, the National Aeronautics and Space Administration, the National Institute of Standards and Technology, the Henry Luce Foundation, and the Doris Duke Foundation, the National Academies assembled an ad hoc committee to examine policies and practices to support family caregivers in STEMM. This expert committee was interdisciplinary and diverse in nature and included a range of individuals who have been nationally and locally recognized for their roles in leading and evaluating effective policies, practices, and programs for supporting family caregivers as well as work-life management in science, engineering, and medicine. The committee included leading scholars and researchers in industrial and organizational psychology, basic science, medicine, human resource management, labor law, economics, sociology, and expertise in federal and state policy.

The committee was charged with detailing current knowledge about the state of family and unpaid caregivers in STEMM careers as well as efforts to support them; to document innovative and promising practices to build

---

BOX 1-1
Statement of Task

An ad hoc committee of the National Academies of Sciences, Engineering, and Medicine will undertake the following set of activities:

1. Summarize the published research on the challenges faced by scientists, engineers, and medical professionals who are family caregivers (i.e., parents and those with eldercare[a] responsibilities, or both), including research on the impact of COVID-19 on these individuals;
2. Document institutional and governmental efforts to support caregivers and the positive and negative impacts of such efforts (if known), including any unintended consequences of well-intentioned policies and practices;
3. Oversee consultant-led, structured interviews with individuals in science, engineering, and medical fields with caregiving responsibilities to understand their needs related to work-life balance and the factors that affect when and if they make use of institutional and governmental policies and resources. This effort will place a particular focus on the experiences of women from multiple marginalized groups (e.g., women of color);
4. Catalogue promising and innovative practices that institutions have used to support family caregivers (which may include those from other sectors), and identify opportunities for greater coordination between government, community, industry, and institutional policies;

COMMITTEE TASK AND APPROACH   15

further support for caregivers in STEMM and identify barriers to effective policy implementation; and to develop a set of consensus recommendations for academic institutions, federal agencies, and other important interested parties to provide sustainable and comprehensive support for scientists, engineers, and medical and health professionals with caregiving responsibilities. The full Statement of Task for the committee is provided in Box 1-1.

In interpreting this Statement of Task, the committee determined its focus would be on academic STEMM and sought to capture the variety of experiences of family caregivers in academic STEMM as well as the wide range of academic STEMM careers. The committee sought to ensure broad coverage of the various groups that make up the academic STEMM ecosystem, including students, postdocs, residents, and other trainees; staff;

---

5. Outline barriers and facilitators to successful implementation of promising practices to support family caregivers, including academic business models, economic trends in the scientific workforce, and the culture and climate in these fields;
6. Summarize what is known about the economic impact of unpaid caregiving performed by women in science, engineering, and medicine, such as gaps in labor force participation, wage inequities, or job/career opportunities;
7. Offer a set of recommendations for how leaders of academic institutions, federal agencies, and others can better support scientists, engineers, and medical professionals with caregiving responsibilities.

Although the primary focus of the study is women caregivers in science, engineering, and medicine, people of all genders, including men, face obstacles as caregivers. Therefore, the study scope will include caregivers of all genders but emphasize women. The study will also take an intersectional approach and place particular emphasis on the experiences of the most marginalized groups in science, engineering, and medicine, such as women of color, who remain particularly underrepresented in these fields. The study will be informed by two public symposia that will be summarized in a proceedings.

---

[a] The committee acknowledges that the term *elder* is considered ageist by many. In recognition of this, the committee has chosen to use the term *older adult* throughout the report. The exceptions to this are in instances where we are directly quoting others or, in this case, quoting the statement of task for the report.

and tenure-track and non-tenure-track faculty in science, engineering, and medicine and nursing fields in universities, academic research centers and institutions, and government national laboratories. Box 1-2 presents an index of the various titles and positions within the STEMM ecosystem.

Many STEMM professionals with family caregiving responsibilities are employed in industry and government and face some of the same challenges outlined above and in the chapters that follow. There is a need to examine the policies, practices, and norms related to caregiving that affect these groups, yet the committee determined that the context for these sectors was substantially different from that of academic STEMM professionals and was beyond the scope of this single report. The committee urges a separate study to address the unique needs and environments of industry and government STEMM sectors. Innovative solutions from each of these workplaces have the potential to stimulate creative approaches to be applied elsewhere. Additionally, caregivers in a wide variety of workplaces encounter challenges and barriers. Though beyond the scope of this report, much can be learned from workplaces outside of STEMM fields. The committee draws on outside examples to generate ideas for creative solutions to support caregivers in Chapter 7.

In recognizing that caregiving takes on many forms, can vary across people, and can change in pace and intensity over time, the committee also chose to adopt an expansive definition to include care for both immediate family and other close individuals; to include individuals of all genders who provide caregiving labor; and to consider both intense, episodic moments of care as well as less intense ongoing caregiving tasks. More detail on the expansive definition of family caregiving employed by the committee can be found in Chapter 2.

In recognition of the many ways that intersecting identities of gender, race/ethnicity, socioeconomic status, and other key demographics shape experiences of both caregiving and academic STEMM and the lack of literature on this topic, the committee adopted an intersectional framework.[1] In doing this, the committee acknowledges that the current literature is underdeveloped in its discussion of the unique challenges that women of color caregivers face in academic STEMM. The committee drew on those

---

[1] *Intersectionality* refers to the interplay of different demographic categories such as race/ethnicity, gender, sexuality, socioeconomic status, and so forth, and how this interplay affects people in unique and nonadditive ways such that the experience of being a Black woman is not simply equal to the experience of being Black and the experience of being a woman. The committee sought to engage an intersectional framework that considers the ways in which caregiving experiences differ among people based on multiple intersecting characteristics, such as highlighting the ways that Black women caregivers, for example, experience certain policies differently than White women caregivers as well as men caregivers.

BOX 1-2
Titles and Positions Within Academic STEMM

The academic science, technology, engineering, mathematics, and medicine (STEMM) workforce is a broad ecosystem composed of individuals occupying a variety of positions, with differing relationships to the university. The list below provides examples of key titles for different groups within this ecosystem but is not intended to be exhaustive.

General STEMM
- Undergraduate student
- Graduate student
- Postdoctoral researcher/postdoc
- Adjunct professor
- Lecturer
- Teaching professor
- Research professor
- Assistant professor
- Associate professor
- Full professor

Medical/Nursing STEMM
- Medical student
- Medical intern
- Resident
- Fellow
- Instructor
- Attending physician
- Nurse
- Nursing student
- Physician's assistant student
- Medical researcher
- Physician

Additional Positions
- Research assistant
- Research associate
- Staff scientist
- Fellow
- Pre-doctoral researcher
- Lab technician
- Teaching assistant
- Intern
- Librarian
- Staff

studies that exist, but more work is needed to provide a more complete and fully intersectional picture. To build its knowledge base for this report, the committee also sought out additional data on the experiences of women of color caregivers in STEMM as part of qualitative interviews conducted for this study, which are discussed later. More specifically, the committee sought to ensure the experiences of individuals with intersecting marginalized identities were not only detailed to describe how experiences of caregiving may vary by identity but were central to considering the ways in which policies and practices should be implemented or approached to be effective for everyone.

ACTIVITIES TO INFORM THE REPORT

To inform its deliberations and findings, the committee engaged in a range of information gathering and research activities. The committee held two national symposia in February and March of 2023. These symposia brought in experts from academia, government, and policy advocacy to discuss key issues related to the challenges family caregivers face in academic STEMM careers. Current federal, state, and institutional policy landscapes supporting family caregivers in academic STEMM careers were presented as well as ideas for possible future landscapes. A proceedings from these events was published in July 2023 detailing each of the talks and panels.[2] The two symposia provided important background for the committee as it began its early discussions and deliberations on the report.

In addition to these symposia, the committee commissioned three papers. Each paper focused on a specific aspect of the Statement of Task. The three papers covered (1) the economic impact of caregiving,[3] (2) current and promising practices to support caregivers in academic STEMM,[4] and (3) challenges faced by caregivers in STEMM and barriers to successful

---

[2] The proceedings of the two symposia, *Barriers, Challenges, and Supports for Family Caregivers in Science, Engineering, and Medicine*, are available at https://nap.nationalacademies.org/catalog/27181.

[3] "The Economic Impacts of Family Caregiving for Women in Academic STEMM: Driving an Evidence-Based Policy Response," by Courtney Harold Van Houtven and Ngoc Dao, is available at https://nap.nationalacademies.org/resource/27416.

[4] "Comprehensive Literature Review of Current and Promising Practices to Support Unpaid Caregivers in Science, Technology, Engineering, and Medical STEMM," by Jessica Lee, Erin Frawley, and Sarah Stoller, is available at https://nap.nationalacademies.org/resource/27416.

policy implementation.[5] These papers provided valuable insights for the committee and formed the backbone to multiple sections and chapters of this report. Instances where the committee drew directly and substantially on these reports are noted at the start of the chapter and in relevant tables.

The committee also sought outside input in the form of a Dear Colleague letter with a call for information on promising practices to support caregivers. This letter sought information from the public on existing efforts to support caregivers at their institutions, particularly novel or unique practices, along with any evaluative evidence published or unpublished on the efficacy of these practices. Responses to this call for information were reviewed by committee members to inform their discussion of report recommendations.

Along with seeking input from the broader scientific community, this report also draws on the results of interviews conducted with caregivers who currently or recently held positions within the academic STEMM ecosystem.[6] The interviews addressed two research questions: (1) How do caregivers' understanding and conceptualizing of factors at the macro level (including structural disadvantage and culture), the meso level (including everyday interactions and social support networks), and the micro level (including personal identities and priorities) shape the ways they engage in and make meaning of caregiving and navigate work-life balance and access policies; and (2) What alternative structures, standards, norms, and supports might better promote work-life balance for caregivers in academic STEMM? Interviewees were asked questions regarding their experiences managing their career and caregiving; cultural, interpersonal, and institutional factors affecting their career and caregiving; ways they might reimagine what success and productivity look like in STEMM; and satisfaction and joy within their career and caregiving. The full interview guide can be found in Appendix B.

These interviews included 40 individuals who within the past 3 years were studying or working at a U.S. university in the sciences, engineering, or medicine and had regular, unpaid caregiving responsibilities of 12 hours or more per week. Interviews lasted approximately 1 hour and followed

---

[5] "A Comprehensive Literature Review of Caregiving Challenges to STEMM Faculty and Institutional Approaches Supporting Caregivers," by Joya Misra, Jennifer Lundquist, and Joanna Riccitelli, is available at https://nap.nationalacademies.org/resource/27416.

[6] "Supporting Caregivers Working in STEMM: Qualitative Study Report," by Tasseli McKay, Monica Sheppard, and Ashley Lowe, is available at https://nap.nationalacademies.org/resource/27416.

a semistructured guide that covered four main topics: (1) experiences of managing career and caregiving responsibilities simultaneously; (2) the macro-, meso-, and microlevel contexts in which caregivers managed those demands; (3) ideas for reimagining success and productivity; and (4) experiences of joy and satisfaction in career and caregiving. Interview recordings were professionally transcribed for analysis. A deductive codebook was developed based on the study research questions and early study committee guidance. Inductive codes were developed jointly by the research team to reflect themes that emerged during the interviews.

In recognition of the many ways that intersecting identities of gender, race/ethnicity, socioeconomic status, and other key demographics shape experiences of both caregiving and academic STEMM and the lack of literature on this topic, the committee specifically sought the experiences of women of color, LGBTQ+ women, immigrant women, and women with disabilities in STEMM as part of the additional interviews conducted for this study. Interview consultants and study staff focused on identifying and connecting with member listservs and similar communication tools that centered these identities. Roughly one-quarter of the sample identified as Black, Hispanic, or Asian; half identified as White; and smaller numbers identified as American Indian or Alaska Native or Native Hawaiian or Pacific Islander. Sixty percent of the sample were from immigrant families, and first-generation college students also represented a majority of the sample. These groups were deliberately oversampled to learn more about their experiences. Interviewees were drawn from across all career stages, from students to senior faculty and academic leadership, with heaviest representation from graduate students, medical residents, and other early-career scholars. More information on the interview sample and analysis can be found in Appendix B.

ORGANIZATION OF THE REPORT

The report responds to the committee's Statement of Task, beginning with an overview of caregiving in Chapter 2. This chapter defines who family caregivers are, details the wide range of tasks and experiences that encompass family caregiving, and outlines trends in family caregiving over the past several decades. Chapter 3 explores the challenges faced by family caregivers in the United States and within academic STEMM with a particular focus on issues of equity and how these challenges are disproportionately borne by certain individuals based on gender and race/ethnicity.

Chapters 4 through 6 engage with the policy landscape of family caregiving. In Chapter 4, the committee outlines existing policies and practices established in federal and state laws, implemented by federal agencies, and developed within individual institutions to support family caregivers. Chapter 5 then examines the barriers to effective policy implementation, highlighting financial, cultural, and practical barriers to policy success. Chapter 6 builds from these two knowledge bases to outline best practices for supporting family caregivers, drawing on existing evaluative research. Chapter 7 considers innovative approaches to flexibility for academic STEMM faculty to push beyond the boundaries of established methods and encourage creative solutions.

The report concludes in Chapter 8 with the committee's recommendations for the ecosystem supporting academic STEMM and the actions that are necessary to support family caregivers and, in doing so, support STEMM innovation and inclusion. This includes recommendations for colleges and universities, federal and private funders, and the federal government. Together, this report represents the committee's expert view on the state of family caregiving in academic STEMM and its hopes for a more inclusive and supportive future for all family caregivers.

2

Overview of Unpaid Family Caregiving

There are only four kinds of people in the world—those that have been caregivers, those that are caregivers, those who will be caregivers, and those who will need caregivers.
– Former First Lady Rosalynn Carter (Carter, 2011)

This chapter provides an overview of national-level data on family caregiving and family caregivers to set the stage for the chapters that follow. It begins by addressing two fundamental questions: who are family caregivers and what is family caregiving? It then turns to examine trends over time in family caregiving, the ways in which the prevalence and demographics of family caregiving have shifted in the past decades, and the impact of family caregiving on caregivers, particularly otherwise employed caregivers. This chapter uses research and data beyond the academic science, technology, engineering, mathematics, and medicine (STEMM) environment to identify trends in how caregiving occurs across workplaces and to provide context for later chapters that examine the specifics of caregiving in academic STEMM. This overview provides necessary background to understand the state of family caregivers to drive action more effectively.

WHAT IS FAMILY CAREGIVING? A TYPOLOGY OF CARE

While the term *caregiving*, encompassing unpaid physical, emotional, organizational, and other support and assistance for loved ones, may be most strongly associated with childcare responsibilities, family caregiving takes many forms. Care for children and young adults is a central aspect of the family caregiving ecosystem, but family caregivers also provide support to aging parents, spouses, and dependent adult children with serious medical conditions, illnesses or disabilities, extended family and kin who may not be blood relations, those approaching the end of life, or those grieving the loss of a loved one. They also manage their own care needs, such as through the use of sick leave to care for their own illness or injury. Each type of caregiving responsibility requires time, attention, energy, and skills, but the configuration of these responsibilities may vary.

As detailed in Figure 2-1, the committee adopted a broad typology of care to capture the various ways in which people across the U.S. academic STEMM workforce provide care for loved ones. Along with childcare and older adult care, which are more commonly understood, spousal care involves care of a spouse or other long-term partner; dependent adult care covers individuals over 18 with physical, mental, or other needs that require support; extended family care includes people outside the nuclear family unit such as kin and community members in need of care; end-of-life care involves the specific support needed for those with a terminal illness; and bereavement care encompasses tasks for those who have lost a close loved one and need assistance. The committee also included caregiver care in this typology to acknowledge the need for caregivers to engage in efforts to ensure they are personally cared for and supported sufficiently to provide care to others. While all these forms of care are encompassed within family caregiving, specific research and data that the committee cites in this report may only speak to one or two forms of caregiving. Given the variation in data, the committee has worked to specify what types of caregiving the data sources are speaking about and worked to identify information on the full range of caregiving included in Figure 2-1.

Each of these types of family caregiving can vary greatly in terms of intensity, duration, and type of support. For example, caring for an adult is different from caring for a child. Caring for adult dependents or aging family members may last for a shorter duration to help manage the effects of a short-term, acute illness or injury or may last for longer due to ongoing disability or deteriorating physical, mental, or cognitive health (Clancy et al., 2020). Additionally, some forms of adult care, particularly that for aging parents, tend to increase over time, while childcare demands often diminish (Duxbury & Dole, 2015). Studies have reported that adult care often can be more complicated to manage with greater unexpected caregiving needs and situations, which may produce greater stress for the caregiver (Smith, 2004). For caregivers of adults, there is the added consequence that policies and programs are often focused on supporting

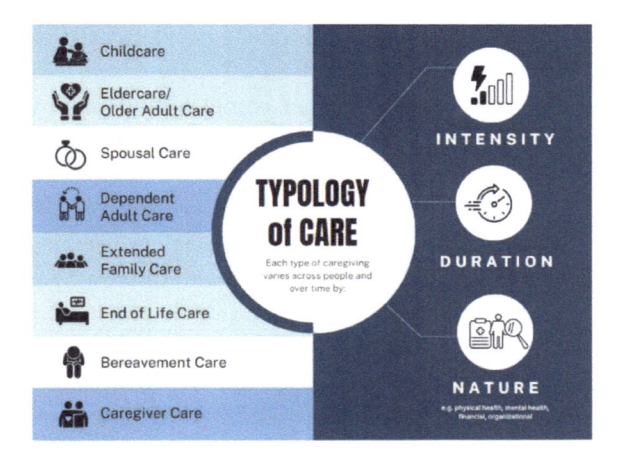

FIGURE 2-1 A typology of care.

childcare responsibilities with less attention to the needs and situations of those caring for older or aging loved ones (Duxbury & Higgins, 2017).

Along with variation across types of caregiving, there is also important variation within each type in terms of intensity, duration, and the nature of care provided both across individuals and over time for an individual caregiver. Those providing care for an adult relative or extended family member may be providing short-term support for physical health needs at one point, then find that as an individual's health declines, there is need for greater-intensity, longer-term support requiring care not only for physical health but also for organizational and financial needs. Caregiving support may address needs related to chronic or acute illness, mental health challenges, and/or physical disabilities. Some research has also distinguished between primary caregivers, those directly responsible for the care of another individual, and secondary caregivers, those who spend less time in direct care, providing instead occasional or less extensive periods of support (National Academies of Sciences, Engineering, and Medicine, 2016).

The amount of time dedicated to caregiving varies depending on several factors, such as the needs of the care recipient; the caregiver's health, work schedule, and professional demands; and the availability of support services. One examination of family caregivers of adults found they spend 23.7 hours per week providing care on average, and the median number of hours spent was 10 hours per week. About 1 in 3 provides care for 21 hours or more each week and 1 in 5 performs 41 or more hours of family care each week, equivalent to the time spent on a full-time, paid job (AARP & National Alliance for Caregiving, 2020).[1]

The caregiving experience is often complex, and family caregivers particularly of adults and children with a serious illness or disability may need to perform a wide range of complex tasks, from medical care to care coordination to technical support. These tasks go beyond the traditional help with activities of daily living, such as bathing and feeding, that have long been the hallmarks of family caregiving (Administration for Community Living, 2022). Of the caregivers of individuals with a chronic illness, disability, or functional limitation, 6 in 10 help with at least one activity of daily living, and 1 in 5 reports difficulty in providing this level of support (Administration for Community Living, 2022). Over time, family caregivers of adults and children with illnesses or disability have reported increasing responsibilities related to activities of daily living, such as preparing meals, managing finances, providing transportation, and administering medications (AARP & National Alliance for Caregiving, 2020). Approximately 70 percent of these family caregivers help monitor the severity of their care recipients' health conditions and nearly two-thirds report that they spend time communicating with health care professionals on behalf of their care recipient (AARP & National Alliance for Caregiving, 2020).

It is challenging for family caregivers to provide this wide array of tasks. Family caregivers may be expected to provide care without a formal assessment of their own needs or the needs of the person they are caring for, and they may be asked to perform tasks that they are not trained in, that they did not expect to have to do, or that they are not comfortable doing. Even when caregivers identify a gap in skills or comfort, they may not be able to easily access training or assistance (Administration for Community Living, 2022). This is not to say that family caregivers provide substandard or inadequate care; on the contrary, much of the research highlights the many benefits of receiving care from a family member or close relation (Callahan et al., 2009; Kokorelias et al., 2019; Samus et al., 2014). Instead, it highlights that this role is complex and can require many different roles

---

[1] Estimates of the amount of time caregivers spent on care also vary greatly across surveys based on differing definitions of caregiving and what is counted as caregiving labor. In a 2019 report from the AARP Public Policy Institute, the authors found survey estimates of average weekly hours ranging from around 5 to over 19. The report cited here examined hours of care per week and stated that caregiving included helping with personal needs, household chores, managing finances, arranging for outside help, and visits to check on well-being (Reinhard et al., 2019).

and responsibilities drawing on many different areas of expertise that may or may not overlap with skill sets family caregivers have already developed.

Ultimately, this variation and breadth of family caregiver roles illustrate that family caregiving, while universal in many ways, does not represent one universal experience. For academic institutions that are committed to supporting their STEMM students and workforce with family caregiving responsibilities, a singular view of family caregivers would be simplistic and problematic and could lead to the development of policies and practices that do not support all caregivers. Understanding the breadth of family caregiver roles and responsibilities and providing the flexibility to engage this diversity is a key component to successful policy implementation.

WHO ARE FAMILY CAREGIVERS?

In the United States, family caregivers are a large and diverse group. National estimates vary,[2] but they indicate that nearly 20 percent of all Americans were engaged in family caregiving for adults 18 and older in 2020 (AARP & National Alliance for Caregiving, 2020),[3] while 40 percent of U.S. families lived with children under 18 in 2022 (U.S. Census Bureau, 2022).[4] While family caregiving touches nearly everyone in some way, the responsibilities of caregiving are not evenly distributed across the population. Women, and particularly women of color, in the United States experience higher societal and familial expectations related to caregiving and, consequently, spend more time providing care than do men. Additionally, caregiving responsibilities were relatively more common among middle-aged workers and those workers who are less educated and with lower income (Cynkar & Mendes, 2011).

Evidence that women disproportionately provide family caregiving support relative to men comes from a variety of rigorous studies and surveys. For example, nationally representative data from the American Time

---

[2] One reason for variation in estimates of family caregivers in the United States is a lack of consistent definitions of caregiving that may included or exclude different activities or differ in the time period over which survey respondents are asked whether they have provided care. In this report, the committee provides what it views as the best current estimates from rigorous reports, but acknowledges estimates are not consistent across the literature.

[3] The AARP and National Alliance for Caregiving report considers those who provided unpaid care to a relative or friend 18 or older at any time over the past 12 months as engaged in caregiving.

[4] These estimates look at all Americans, not only the population of employed Americans.

Use Survey consistently show women's deeper engagement than men's in housework and childcare, although over time, men have increased their involvement in these activities (Bianchi et al., 2006; Sayer, 2005; Wang & Bianchi, 2009). The Health and Retirement Study, another nationally representative sample, reported that gendered patterns in time spent on caregiving continue as people age and extend beyond caring for children (Lee & Tang, 2015). At the same time, even when women do not have caregiving responsibilities, research has found that people may make assumptions that women will become mothers, regardless of their stated intent, and thus will need to take on this responsibility (Thébaud & Taylor, 2021). And even in the paid workplace, women also frequently take on caregiving-type labor, such as mentoring or providing emotional support to colleagues and filling in when others are sick or need to step away due to outside responsibilities (Misra et al., 2021; Moore, 2017; O'Meara, 2016; Writer & Watson, 2019).

While research has shown an increasing move toward greater sharing of childcare responsibilities by men, a similar move in the direction of gender parity has not been observed in older adult care, including care for a woman's husband's parents (Grigoryeva, 2017). Women are more likely than men to be the sole caregiver or provide the majority of care for an adult family member and to provide a greater intensity of care than men (AARP & National Alliance for Caregiving, 2020). Data from the Health and Retirement Study show that women are more likely to provide care for parents and for grandchildren than are men (Lee & Tang, 2015). Daughters are more likely to provide care for older parents than are sons, with sons providing even less care when they have sisters (and therefore daughters providing more care when they have brothers) (Grigoryeva, 2017). Increasingly as well, many women are finding themselves providing what has been termed *sandwich care*, or care for both young children and adult dependents/aging relatives or kin (Pierret, 2006; Suh, 2016).

Compared with the extent of literature on gender disparities, the literature on racial disparities—and particularly examinations of disparities based on intersecting marginalized identities—is limited.[5] While the body of evidence is less robust, the existing data suggest that racial disparities also exist among who is most likely to be providing caregiving, particularly among

---

[5] The limitations of the current literature on caregiving and race/ethnicity additionally do not allow the committee to tease out further differences within groups based on other intersecting identities, such as immigrant status, socioeconomic status, and sexuality.

those caring for adults. For instance, a recent study found that Black and Hispanic women, on average, perform higher levels of intensity of adult care compared with White caregivers, spending nearly 30 hours more per month than their White counterparts on adult care (Cohen et al., 2019). Estimates on time spent on care by race/ethnicity vary, however. Another study found that Black caregivers spent an additional 13 hours a week on care-related tasks compared with White caregivers (McCann et al., 2000). And a recent report noted that Black caregivers spend around 10 hours more per week and Hispanic caregivers around 5 hours more per week on caregiving than White caregivers (AARP & National Alliance for Caregiving, 2020). These disparities have also been found to hold in academic STEMM as well. For example, one survey carried out at a research-intensive public institution in the Northeast, which looked at the relationship between working hours and care work outside of the university for all faculty, found that faculty of color spend more time than White faculty on older adult and long-term care (Misra et al., 2012). Black caregivers are also more likely to provide informal care beyond immediate family members to others, such as friends or church members (Cohen et al., 2019; McCann et al., 2000). Historic distrust in caregiving institutions and medical institutions due to histories and continuing experiences of racism and discrimination may contribute to greater reliance on informal care among marginalized groups (Dilworth-Anderson et al., 2020).

Relatedly, Asian American and immigrant caregivers often encounter different cultural expectations for care than do many White Americans. Expectations of filial obligation to care for older adults and aging relatives are particularly strong among many Asian caregivers and influence decisions about providing care personally versus seeking the support of paid caregivers (Guo et al., 2019; Moon et al., 2020). There are also stronger expectations of grandparents providing care to grandchildren, which not only has been shown to have positive effects on mental health but also influences the caregiving burden of many Asian grandparents (Xu et al., 2017). In this way, expectations of providing unpaid caregiving can look different from presumed norms of White caregivers.

Research on caregiving demands among American Indian and Alaska Native populations has especially been lacking; however, one recent survey of 225 participants found that 40 percent of respondents had provided unpaid care over the course of at least 1 month. Respondents provided care for children and adult relatives as well as friends and other loved ones. While the authors noted the role of strong cultural ideologies of community responsibility played a role in the prevalence of informal caregiving, they also hypothesized that this view of caregiving as a duty and not a burden may provide a buffer against caregiving stress and help to explain the high degree of satisfaction in providing caregiving assistance among respondents (Strachan & Buchwald, 2023). More research, however, is needed on the experiences of both giving and receiving care among American Indian and Alaska Natives.

Overall, research demonstrates that, although there is no one type of person who is a family caregiver, caregiving falls disproportionately to women. Women of color may also face particular challenges given intersecting biases of gender and race/ethnicity, though more work is needed to tease out intersectional differences. Along with variation by race/ethnicity and gender, within the academic STEMM community, caregivers are present at all levels. For instance, caregivers represent a significant portion of the STEMM student and trainee population. More than 1 in 5 undergraduate students are parents (Gault et al., 2020). Whether engaged in the paid labor force, earning a degree, or both, more and more individuals over time are balancing the challenges of caregiving with the challenges of employment in academia and education.

RISING NATIONAL TRENDS IN FAMILY CAREGIVING NEEDS

Over the last several decades in the United States, significant changes have occurred in demographic characteristics of the population, the composition of the professional workforce, and in societal norms and patterns that have far-reaching implications for family caregiving.

Increase in Need for Care

Significant shifts have occurred in the past several decades that have increased the U.S. population in need of care. The U.S. population has aged, particularly between 2010 and 2020, when the country experienced the largest increase in the population 65 and older since the late 1800s (Caplan & Rabe, 2023). According to the Administration for Community Living, the number of Americans aged 65 and older increased 18-fold, from 3.1 million in 1900 to 55.7 million as of 2021, and today's Americans are living nearly 30 years longer than their 1900 counterparts (Administration for Community Living, 2022). As a result, more of the U.S. population is likely to need care from a family member, friend, or direct care worker than in previous periods (Administration for Community Living, 2022). Moreover, during the last two decades, the United States waged its longest war, and due to enhanced medical care on or near the battlefield, many veterans survived with life-altering injuries that continue to require long-term or lifetime care, often provided by family members (Bilmes, 2021). Even more recently, the COVID-19 pandemic may further affect societal needs for caregiving related to long COVID and other forms of morbidity (Boyd et al., 2022; Isasi et al., 2021).

Regarding childcare, however, births in the United States have decreased. In 2022, the U.S. Census Bureau reported that there were 63 million parents with children under the age of 18 living in their home, a 5 percent decrease from 2010 (U.S. Census Bureau, 2022). Despite this shift, which has happened at the same time as increases in workforce participation among women, the amount of time that parents spent providing active care to their children has risen over several decades (Bianchi et al., 2006; Sayer et al., 2004). This is especially true for highly educated parents, such as many in the academic STEMM workforce. Some research has found that these parents spend more time providing active care to children than those with less education (England & Srivastava, 2013). In addition, nearly 3 million grandparents are primary caregivers to millions of children who cannot remain with their parents (Administration for Community Living, 2022).

The increase, notably in the 1970s, of women, particularly college-educated, White women, entering the paid labor force in record numbers and remaining employed after the birth of their children is also a significant cultural and social factor that explains in part the increased need for caregiving support. This shift in employment dramatically altered women's past roles as housewives, mothers, and daughters who were available to provide full-time care for children and adult family members (Donnelly et al., 2016; Goldin, 2023).

Increase in People Providing Unpaid Care

A significant increase in the number of people who provide unpaid family care has been occurring for quite some time (Kossek, 2006). For example, in 2020, an estimated 53 million adults provided unpaid care to either an adult or a child with special needs, up from 43.5 million in 2015. Thus, more than 1 in 5 Americans are now caregivers for adults or for children with special needs (AARP & National Alliance for Caregiving, 2020). In fact, "since the 1990's (when statistical tracking began), the United States

(US) has seen growth in the number of people engaged in family caregiving, the number of weekly hours they provide assistance, the difficulty of their caregiving tasks, and their labor force participation rate" (Lerner, 2022).

While the greater need for caregiving has played a substantial role in the increase in the number of people engaged in unpaid family caregiving, this increase in unpaid caregiving is likely due to several additional, large-scale cultural and economic factors. These include the increased financial costs of professional caregiving,[6] increased recognition of what counts as caregiving, the increase in women's participation in the paid labor force, and an insufficient labor force of paid professional caregivers.

While the overall population needing care has grown, the rising cost of long-term care has made it difficult for many families to afford professional care (Abelson & Rau, 2023; Administration for Community Living, 2022). Already, caregiving is incredibly costly, as many family caregivers incur large out-of-pocket expenses in the thousands of dollars or more to care for adult loved ones (AARP, 2021) and hundreds of thousands of dollars to raise a child to age 18 (LaPonsie, 2022; Lino et al., 2017). Adding the cost of paid care services can be too great an additional burden and many families instead have chosen to take on the responsibility of care themselves (AARP & National Alliance for Caregiving, 2020).

We may also attribute some of the noted rise of caregiving in the United States to increases in the number of self-reported family caregivers in national surveys. In recent years, as greater attention has been paid to family caregiving responsibilities in the media, more people recognize that the everyday support they provide to their family and other loved ones is a form of caregiving, which may result in a higher proportion of people self-identifying as caregivers on national surveys (AARP & National Alliance for Caregiving, 2020).

Within that same period that has seen an increase in family caregivers, there has also been a simultaneous increase in the paid labor force participation of family caregivers. This is in part due to the shift in the 1970s when more women entered the paid labor force and thus were faced with combining their paid work with unpaid caregiving responsibilities. Today, a

---

[6] Recent estimates suggest the cost of paid childcare increased 86 percent between 1995 and 2016, and the cost for long-term care for older adults has also increased as the demand for such care has risen with an aging population (Hayes & Kurtovic, 2020; Swenson & Simms, 2021).

very large share of individuals combine their family caregiving responsibilities with paid work. According to Current Population Survey data, nearly 73 percent of all mothers with children under age 18 and 93 percent of all fathers with children under age 18 were in the labor force in 2022 (U.S. Bureau of Labor Statistics, 2023b). More than 80 percent of employed mothers work full-time, and full-time work is nearly universal among employed fathers (95.6 percent) (U.S. Bureau of Labor Statistics, 2023b). Additionally, according to a 2010–2011 U.S. Gallup Survey of a random sample of nearly 250,000 individuals ages 18 and above, 1 in 6 respondents who reported working a full- or part-time job also reported assisting with care for an older or disabled family member, relative, or friend (Cynkar & Mendes, 2011).

Finally, the increase in reliance on unpaid caregivers, especially of those who also work, is, in part, due to the scarcity of professional caregivers available to hire. For decades, scholars have detailed the gap between the supply of paid caregivers and the demand for their work (Super, 2002). Recent estimates project a national shortage in caregivers for older adults of 151,000 by 2030 and 355,000 by 2040 (Global Coalition on Aging & Home Instead, 2021). Not only does this lack of professional caregivers result in the caregiving needs falling to unpaid family caregivers, but it can also lead to the unpaid caregivers spending significant time working to find and hire a professional caregiver, which on its own is a form of providing care. In 2020, about one-third (31 percent) of family caregivers for adults or children with special needs experienced at least some difficulty in coordinating care for their loved one, up from 23 percent in 2015 (AARP & National Alliance for Caregiving, 2020). While this report focuses on the experiences and needs of unpaid family caregivers, it should also be acknowledged that greater support for paid caregivers is important to the success and support of unpaid caregivers.

Increase in Intensity of Care During the COVID-19 Pandemic

The intensity of care, defined as both the number of hours in a week required to provide care to a care recipient and the difficulty and complexity in the types of tasks caregivers are required to perform, was particularly notable during the COVID-19 pandemic. The pandemic played a significant role in shaping caregiving intensity, as many care centers closed and family caregivers worked to balance the unique challenges of work during a pandemic with caregiving needs. Studies examining the experiences of both those caring for children and those caring for adults reported increases in care intensity and burden of care during the pandemic (Archer et al., 2021; Cohen et al., 2021). This was particularly acute for women who traditionally and still today often take on the greatest share of family caregiving labor (Cohen et al., 2021). The pandemic not only contributed to increased intensity of care, but it also increased risk of health-related socioeconomic vulnerabilities for caregivers more than for non-caregivers. The pandemic contributed to worsened financial strain, food insecurity, housing insecurity, interpersonal violence, and transportation difficulties for family caregivers (Boyd et al., 2022).

THE IMPACT OF CAREGIVING AND THE CHALLENGES FACED BY CAREGIVERS

Caregiving across the life course is an important and rewarding role for many and has a deeply personal effect on all who need care. Research is clear that caregiving experiences in early childhood are formative for later well-being, and that care quality has a substantial influence on the experience of aging and/or disability particularly by allowing adults needing care to remain in their homes (Fernandez et al., 2016; Worthman et al., 2010). Parenting; spending time caring for aging parents; and connecting to other family, friends, and communities through providing care all give individuals opportunities for relationship building and nurturing (Mackenzie & Greenwood, 2012). Nevertheless, research is also clear that family caregivers often experience physical, emotional, and financial challenges, which indicates a growing need for support services for them.

As family caregiving can in many instances involve high intensity and complexity, it has associated effects on the well-being of family caregivers (AARP & National Alliance for Caregiving, 2020). AARP and the National Alliance for Caregiving (2020) define "intensity of care" by the number of hours of care given as well as the number of activities of daily living a caregiver assists with (e.g., bathing, toileting, feeding). Approximately 40 percent of unpaid family caregivers for adults or children with special needs are in high-intensity caregiving situations, with 16 percent experiencing a medium intensity and 43 percent experiencing a low intensity. High-intensity caregiving is often associated with worse self-reported health outcomes and higher rates of financial strain (AARP & National Alliance for Caregiving, 2020).

During the COVID-19 pandemic, which exacerbated demands for unpaid care across the life course, individuals with caregiving responsibilities experienced negative mental health outcomes at much higher rates than non-caregivers. Approximately 70 percent of those parenting children or caring for adults reported adverse mental health symptoms, including anxiety or depression (Czeisler et al., 2021). The consequences of caring for adults in this challenging environment was more significant than that of caring for children, and those who reported the highest rates of negative mental health effects were those who were caring for both children and adults. This last group was almost four times more likely than non-caregivers to experience adverse mental health symptoms (Czeisler et al., 2021). Research has also shown that particularly for women, a lack of support for caregiving and the increased demands during the pandemic resulted in higher levels of psychological distress, understood as experiences of anxiety, worry, depression, and hopelessness (Prados & Zamarro, 2020; Ruppanner et al., 2019; Zamarro & Prados, 2021). As workplaces deal with the mental health impacts of COVID-19, it is important that institutions attend specifically to caregivers and understand that the negative effects of caregiving on mental health preceded the crisis.

Caregiving also affects labor force participation among caregivers, particularly for mothers of young children (Cortés & Pan, 2020), which has financial consequences. According to surveys conducted by Gallup-Healthways and Pfizer-ReACT, Americans who identify as caregivers working at least 15 hours a week miss an average of 6.6 workdays per year. This absenteeism results in 126 million missed workdays a year, which would be even higher if part-time workers were included in these calculations (Witters, 2011). As a result of hours spent on care, many caregivers leave paid jobs, cut back on work hours, miss days at work, or limit funds put into their retirement savings (AARP & National Alliance for Caregiving, 2020; Weller & Tolson, 2019; Witters, 2011). These reduced or lost earnings can slow their wage growth over time and ultimately limit their retirement income from Social Security and employer-based retirement plans (Weller & Tolson, 2019). Table 2-1 provides an overview of the causal evidence on the economic impacts of caregiving for employed family caregivers.

Overall, the evidence suggests mothers in particular face greater challenges following the birth of a child, especially a first child, as well as clear penalties in job opportunities and earnings and reductions in productivity. Fathers, instead, do not face these same penalties and challenges to the same

TABLE 2-1 Causal Effects of Caregiving on Key Economic Outcomes

Outcome	General Findings: Caregivers of Children	General Findings: Caregivers of Adults
Labor Force Participation	Mothers, but not fathers, are more likely to exit the labor force after becoming parents (Cortés & Pan, 2020). The strongest impact is after the birth of a first child (Doren, 2019).	Most of the research is descriptive, finding largely modest or null, but negative, effects (Aughinbaugh & Woods, 2021; Fahle & McGarry, 2018; Reinhard et al., 2023; Wilcox & Sahni, 2022). The few causal studies present mixed findings with either modest negative effects (Jacobs et al., 2016; Maestas et al., 2023) or null effects (Stern, 1995; Van Houtven et al., 2013). Research also suggests caregivers are more likely to retire, but the effects are small (Jacobs et al., 2016; Miller, 2009; Van Houtven et al., 2013).
Hours of Paid Work	There is limited causal research on the impact of children on paid hours worked among STEMM professionals in the U.S. context. Evidence from other fields, such as law and business, as well as international evidence finds a decline in women's hours worked following the birth of a first child (Azmat & Ferrer, 2017; Bertrand et al., 2010; Kleven et al., 2019) and an increase in part-time work (Boelmann et al., 2021; Schmitt & Auspurg, 2022).	Caregivers of adults are more likely to work fewer paid hours than non-caregivers, but the effect size is small (He & McHenry, 2015; Johnson & Lo Sasso, 2000; Van Houtven et al., 2013).
Job/Career Opportunity	Lab, audit, and quasi-experimental studies find that mothers are discriminated against in hiring (Correll et al., 2007).	There is limited research on the causal connection between how caregiving of adults affects job and career opportunities, and thus no finding about the effect can be identified.

TABLE 2-1 Continued

Outcome	General Findings: Caregivers of Children	General Findings: Caregivers of Adults
Earnings/ Wage Penalties	Women, particularly married women, younger women, and women of color, face a motherhood wage penalty of around 4–7% (Anderson et al., 2002; Budig & England, 2001; Kahn et al., 2014). Other research has found that these effects may grow over time, with estimates of long-term wage penalties in the United States at 31% (Kleven et al., 2019).	Studies have provided mixed conclusions, ranging from no effect for men or women (Van Houtven et al., 2013) to no effect for men but a small effect for women (Butrica & Karamcheva, 2015) to large effects for women (Nizalova, 2012; Skira, 2015), particularly younger women (Maestas et al., 2023).
Productivity	Without adequate support, mothers in particular experience penalties to their productivity given competing demands (Morgan et al., 2021). COVID-19 and the subsequent challenges with childcare coupled with ineffective organizational support also resulted in productivity gaps between men and women (Kossek, Dumas, et al., 2021; Stall et al., 2023).	Only one study was identified on the association between productivity and caregiving for adults. This study found a substantial decrease in work productivity due to caregiving demands (Mazanec et al., 2011).

NOTE: This table draws substantially from the research paper "The Economic Impacts of Family Caregiving for Women in Academic STEMM: Driving an Evidence-Based Policy Response," by Courtney Van Houtven, Ph.D., and Ngoc Dao, Ph.D., that was commissioned for this study. The full paper is available at https://nap.nationalacademies.org/resource/27416. More on the methodology for inclusion of studies in this paper can be found in Appendix C.

degree. The findings for caregivers of older adults are less conclusive and generally present mixed, null, or small effects on the outcomes examined.

Importantly, however, all these effects are not simply the inevitable outcome of a need to provide care for another person. Instead, many of them are the result of a lack of support for caregiving and could be mitigated with greater support. For example, access to affordable childcare is one key intervention that has been shown to have an immense effect on not only the likelihood of being employed but also the hours spent in paid employment among American women (Ruppanner et al., 2019). Greater support from partners also makes a key difference in employment outcomes for mothers. Research conducted during the pandemic found that employed mothers with less support from their partners reported a greater reduction in their working hours than those with more support (Prados & Zamarro, 2020).

For caregivers who are enrolled as students, there are also financial challenges stemming from a lack of support for caregiving. Most of the existing data on the challenges of student caregivers focus on student parents. These students are significantly less likely to graduate within 6 years: while nearly 60 percent of all students graduate in 6 years, under 40 percent of student parents graduate in that time frame (Institute for Women's Policy Research & Aspen Institute, 2019). Students with children also have higher rates of educational debt than students without children, as they often face greater financial responsibilities as well as insecurity and are more likely to be enrolled in for-profit institutions (Institute for Women's Policy Research & Aspen Institute, 2019). All of this contributes to greater financial burdens associated with both their caregiver status and educational status for student parents.

Challenges faced by family caregivers are exacerbated in the U.S. context compared with other developed countries because of the lack of strong public policy supporting unpaid caregivers, leaving more of the burden of addressing the needs of caregiver-workers on individual employers (AARP, 2021; Body, 2020). Of the high-income countries in the Organisation for Economic Co-operation and Development (OECD), the United States is the only one that lacks national statutory paid leave for parents and other unpaid caregivers (Gromada & Richardson, 2021). Combined with its poor investment in paid childcare, this distinction places the childcare system in the United States 40th overall out of the 41 high-income OECD countries (Gromada & Richardson, 2021). In health care, the United States has the highest levels of spending and the worst health outcomes of any wealthy nation (Gunja et al., 2022). The absence of a national health care system that guarantees care to all is particularly notable. Care for those who are aging or disabled is similarly fragmented and inadequate.

These challenges were poignantly displayed among interview participants. Interviewees noted expending tremendous intellectual, financial, and physical resources in the attempt to manage the competing demands of their careers and caregiving responsibilities. They described distilling their priorities and triaging their workloads; devising new time-efficiency strategies; learning to fit in more flexible work and caregiving commitments around inflexible ones; communicating proactively with advisors, teachers, managers, and academic leadership; and developing a host of creative personal and professional arrangements to attempt to fulfill their competing academic and professional responsibilities. Such strategies were often a source of pride, and some success.

Caregivers with access to substantial private resources, such as individuals in academic leadership roles or those in dual-career physician couples, recounted using every resource, unpaid and paid, at their disposal to make their professional situation tenable in the context of substantial caregiving commitments. As one respondent stated:

> "The only way I was able to make it work was my husband was a stay-at-home dad at that point.... He'd get up in the middle of the night with changing my parents' bedclothes if there were accidents and things like that.... The only way we were able to make that work was him being at home. And I think that's a real problem because not everybody has that flexibility. I probably would've had to put my parents in a nursing home if it wasn't for that [or] I would've had to quit my job ... the combination of financial resources and partner resources helped me to care for kids and parents at the same time.... I wouldn't have survived without that."

Those who had fewer private resources, particularly students and other early-career scholars and those from underrepresented backgrounds, more often recounted being driven out of their professions or scientific careers entirely. Others left academia for jobs in industry. One interviewee discussed her own thoughts on leaving the academy:

> "Not everybody comes from a privileged background. So those expectations that you have to work for these high-risk, high-reward projects for many years and put the rest of your life on hold, they are nonsustainable for most of us, right? And especially if you have a young family. I mean, you can decide to sacrifice your time and yourself, but you cannot do that for your family."

Still, most interviewees—even those occupying positions of relative personal or professional privilege—experienced the conflict between their careers and their caregiving responsibilities as irreconcilable. As one interviewee noted:

> "In a perfect world, I could balance it all. I could just be really efficient at work, get through my calls, get through my notes, then come home and have dedicated time to spend caring for my mom. But the time just doesn't allow for it. There's often times where I'm staying late trying to catch up ... it just squeezes how much care I can do.... You can't be great at either [career or caregiving] at the same time."

SUMMARY OF FINDINGS FROM CHAPTER 2

Demands for caregiving have risen simultaneously with demands on the time of family caregivers in paid labor, resulting in greater binds and constraints for today's family caregivers. Gender and racial disparities in caregiving roles and responsibilities mean that, in general, these challenges are experienced disproportionately by women and may be especially acute for women of color.

1. Family caregiving takes many forms, including care for children and young adults (those with and without special needs), aging relatives, spouses, dependent adult children, extended family and kin who may not be blood relations, those approaching the end of life, and/or those grieving the loss of a loved one. Each of these types of caregiving can vary in terms of intensity, duration, and type of care provided both across individuals and over time for an individual caregiver.
2. Family caregivers in the United States are diverse, with caregivers coming from all backgrounds and demographics; however, these demands are not evenly distributed across the population. Women are more likely than men to care for children, older parents, and grandchildren, and women of color are more likely than White men or women to provide care for extended family, such as siblings, parents, and care for kin who may not be blood relatives.
3. Since the 1990s, the number of people engaged in caregiving, weekly time spent providing care, and the labor force participation rate of caregivers have all increased. These increases are likely due to several factors, including large-scale cultural and economic

shifts; growing aging, chronically ill, and disabled populations; increased self-reporting of caregiving roles in national surveys; increased costs for professional caregivers; and an insufficient supply of paid professional caregivers.

4. Caregiving across the life course is an important and rewarding role for many, and its effect is significant—research is clear that caregiving experiences in early childhood are formative for later well-being and that care quality has a substantial effect on the experience of aging and/or disability. Nevertheless, research also demonstrates that family caregivers often experience physical, emotional, and financial challenges, which were exacerbated during the COVID-19 pandemic as well as by insufficient national- and institutional-level support.

5. Individuals across academic STEMM face challenges due to a lack of support for caregiving. Certain populations, particularly students and trainees, who are earlier in their career, less established, and in more precarious positions, may face particular challenges.

3

Caregiving Challenges and Implications for Equity in STEMM

This chapter draws substantially from the research paper "A Comprehensive Literature Review of Caregiving Challenges to STEMM Faculty and Institutional Approaches Supporting Caregivers," by Joya Misra, Ph.D., Jennifer Lundquist, Ph.D., and Joanna Riccitelli, which was commissioned for this study.[1]

As greater attention to caregiving is needed across the labor force given the experiences and trends detailed in the last chapter, the conflicts between family caregiving and paid labor are particularly acute for science, technology, engineering, mathematics, and medicine (STEMM) fields. This chapter details the ways in which assumptions about gender and race/ethnicity intersect with cultural norms and rewards systems at work to produce challenges for family caregivers in academic STEMM. It describes how family caregiving responsibilities clash with ingrained norms in academic STEMM fields that implicitly assume that all STEMM workers and students can exhibit unwavering devotion to STEMM and remain constantly available and visible when learning and working in these fields. These norms are also buffered by systems that reward those who work long hours through grants, promotions, tenure, and raises. Such norms hurt creativity and innovation in STEMM, drive bias and discrimination against caregivers who may be seen as unable to meet these norms regardless of actual productivity, and ultimately affect the structure of learning, working, and advancement in STEMM in ways that fundamentally undermine the retention and advancement of family caregivers in academic STEMM. People of all genders who have caregiving responsibilities are negatively affected by these norms, but they have a disproportionate effect on women who are most often in caregiving roles. Unpacking the ramifications of "ideal worker" norms and

---

[1] The full paper is available at https://nap.nationalacademies.org/resource/27416.

flexibility stigma within academic STEMM is crucial to understanding the cultural context that both shapes current policies and practices and directs attention to where action is needed to promote change.

IDEAL WORKER NORM AND WORK DEVOTION SCHEMA[2] IN STEMM

In academic STEMM, strong and resilient norms about ideal workers shape our cultural conceptions of how people should approach working. These norms amplify tensions with any responsibilities in a person's life outside of paid labor (Blair-Loy, 2001; Kossek, Perrigino, & Rock, 2021; Williams, 1989). As described by Kachchaf and colleagues (2015), the ideal worker in STEMM is characterized by "commitment to the job through long hours, unbroken career trajectories, and constant availability and visibility." The ideal worker norm assumes that the individual worker has no significant care responsibilities and, instead, is cared for by other members of their household (Kachchaf et al., 2015). Ideal worker norms in STEMM reflect the implicit assumption that STEMM students and professionals are White, heterosexual, upper-class men without caregiving responsibilities (Acker, 2006; Bird, 2011; Carrigan et al., 2011; Gatta & Roos, 2004; Kachchaf et al., 2015; Sallee, 2012). As Gatta and Roos (2004, p. 124) argue, the assumption underlying the ideal worker norm is that all workers have a "full-time wife at home fulfilling the roles of childcare worker, eldercare provider, maid, launderer, and chef, among other duties."

Ideal worker norms produce a set of productivity standards in STEMM that reinforce expectations of working around the clock with little recognition of outside needs or even the physical limitations of workers (Drago et al., 2006; Ecklund & Lincoln, 2016; Kachchaf et al., 2015; Sallee, 2012). Academic organizations have earned a reputation for being "greedy institutions" that expect STEMM academics to spend substantial and intensive hours working (Cech & Blair-Loy, 2019; Ecklund & Lincoln, 2016; Kachchaf et al., 2015; Kossek & Lee, 2022; Misra et al., 2012; Sallee, 2012; Ward & Wolf-Wendel, 2012). Researchers note that many STEMM faculty are expected to work "60–80 hours per week, without time constraints or boundaries," making balancing caregiving and career challenging (Ecklund

---

[2] Cultural schemas are cognitive patterns individuals use to organize, understand, and interpret their social world. They help to create meaning and can shape how people act and interact.

et al., 2017). In the words of one participant in a qualitative study of how women faculty with small children manage their parental and professional roles at research universities (Ward & Wolf-Wendel, 2012):

> "The biggest thing for me is that I feel like I don't have time. I used to work so many more hours and I just don't have those hours anymore. And I'm constantly struggling . . . I mean, during those hours I feel I have so much to do. But I don't get the time to stop and think and do creative research, I'm just kind of up-keeping all the time."

Importantly as well, long hours are frequently rewarded with higher pay, producing even greater pressure to work longer (Cha & Weeden, 2014; Goldin, 2023). All of this contributes to challenges for many individuals to set boundaries for themselves and freely choose their time spent on work, as rewards for meeting these expectations and censure for failure to do so are high.

There is a moral component to long hours within academic STEMM, as the culture frequently demands intense devotion to work based on the belief that an individual's work is not simply a job, but instead a professional identity and vocation (Blair-Loy & Cech, 2022). Workers who accept a work-devotion schema as true embody the following:

> "A cognitive acceptance of the legitimacy or intractability of work demands, a moral and emotional identification with one's employer or profession, inspiration and transcendence of personal limitations from the projects and relationships that work provides" (Blair-Loy & Cech, 2017).

While these culturally reinforced norms and rules may seem most explicitly to apply to university employees, they also affect students, residents, fellows, and even hourly workers who are building toward academic and research careers. Cultural expectations for work devotion and professional sacrifice are particular challenges for these groups because they lack the status, power, and autonomy to set firm boundaries when faced with expectations to work long hours (Lambert et al., 2022).

The ideal worker norm and work devotion schema in STEMM originate from a time when these fields were less diverse than they are today. While many STEMM fields still struggle to be diverse and inclusive, due in large part to the challenges outlined in this report, the demographic makeup of STEMM education and the workforce has undergone a major shift over the last several decades. Women are now in the majority of most medical school classes and are at parity in many fields in the life sciences (National Academies of Sciences, Engineering, and Medicine, 2020). The number of STEMM degrees earned by women of color has doubled over the past 10 years (National Center for Science and Engineering Statistics, 2021). Social norms about fatherhood are also changing in ways that conflict with ideal worker norm expectations in STEMM. Today, faculty men are experiencing both greater desire and greater pressure to contribute to family caregiving (Damaske et al., 2014; Marotte et al., 2011). Ideal worker and work devotion mandates are incompatible with today's more diverse STEMM workforce, create substantial barriers to participation of family caregivers in STEMM, and thereby undermine the vitality of the STEMM workforce.

CONSEQUENCES OF IDEAL WORKER AND WORK DEVOTION NORMS TO STEMM INNOVATION AND VITALITY

Research suggests that ideal worker and work devotion norms may serve to stymie innovation in STEMM, drive potential burnout, and reinforce outdated assumptions about gender. Constant work can serve to diminish creativity and discovery. Recent psychological literature has shown that time for rest and to allow one's mind to wander rather than remaining intently focused on one task or goal provides greater space for discovery and creativity (Newport, 2016; Pang, 2016). Current work devotion schemas, however, do not provide much space for the kind of rest and time away that can be so fruitful for innovation (Blair-Loy & Cech, 2022).

Furthermore, overwork driven by these norms can lead to burnout, which results from unrelenting workplace stress (World Health Organization, 2019). In the most stressful overwork environments, STEMM workers may risk significant individual consequences on health as well as experience greater intentions to drop out, decreased job satisfaction, and less connection to their work (Mayo Clinic, 2023; National Academies of Sciences, Engineering, and Medicine, 2019; National Academy of Medicine, 2022). One examination of burnout among women faculty in computer science departments found that women were more likely than men to consider leaving during the pandemic because of increased work-family conflict and burnout, along with decreased job satisfaction (Lawson et al., 2023).

Ideal Worker Norms and Gender

Ideal worker norms are harmful to all groups, but they have different consequences for men than they do for women. For women, they fuel bias—assumptions that women are not ideal workers because they have responsibilities to care for others outside of work and therefore are not qualified for professorial positions or for promotions (Williams, 2000). For men, they fuel pressures to work the same kind of hours and patterns as someone who has no ongoing responsibilities for daily care of others (Williams, 2000) and limit their ability to be involved in caregiving (Reddick et al., 2012; Sallee, 2012; Sallee & Lester, 2009; Sallee et al., 2016). The gendered societal expectation that women will carry out a larger share of household labor and caregiving means that a man's desire to be an involved caregiver is seen as out of the ordinary, and perhaps an invalid reason for placing boundaries around their time at work (Gheyoh Ndzi, 2023). And for women, these expectations result in more permeable work-nonwork boundaries and often more work-life conflict due to increased role overload from trying to carry out caregiving and work demands at the same time (Kossek & Lee, 2020, 2022). One consequence of ideal worker norms is that academic mothers and fathers are often reluctant to use supportive policies, such as stop-the-clock tenure policies, because doing so could be seen as violating the ideal worker norm that they always should be working and that familial obligations either do not exist or, if they do, do not infringe upon availability or productivity (Drago et al., 2006; Williams & Lee, 2016). Sallee (2012) and colleagues (Sallee et al., 2016) find that gendered norms for parenting led to male faculty perceiving or being explicitly told that they were unable to avail themselves of stop-the-clock policies and other accommodations intended to support faculty parents following the birth or adoption of a child. This highlights how policy changes to facilitate the balance of caregiving and work are insufficient if they do not address the cultural barriers that constrain who is seen as a valid candidate for such policies and who is not (Kossek et al., 2009).

Given all of this, ideal worker norms harm those, regardless of gender, who engage in family caregiving because this work challenges assumptions of complete devotion to work. This custom operates differently for men and women but has consequences for both. Additionally, ideal worker norms produce unique gender consequences for women that they do not for men. Along with facing challenges when they are in fact providing caregiving labor outside of paid work, women also face assumptions that they will inevitably become mothers or caregivers and thus will at some point become less committed to their work (Thébaud and Taylor, 2021). Given this, women face potential bias as caregivers even when they are not. Both of these factors can be at play and influence how caregivers, women in general, and particularly women caregivers are treated and evaluated in the workplace.

BIAS AND DISCRIMINATION AGAINST CAREGIVERS

Maternal Wall Bias and the Motherhood Penalty

Ideal worker norms in STEMM fields give rise to bias and discrimination against caregivers, particularly women. For example, a large body of research has documented a "motherhood penalty," which is defined as negative effects on pay, hiring, and advancement for mothers whether through discrimination or other factors, as well as "maternal wall bias," which is defined as discrimination against mothers simply for being mothers, across industries and occupations. Both of these barriers operate in part based on beliefs that mothers are less committed to the organization, and even less competent, than women without children and men with and without children (Benard et al., 2004; Correll et al., 2007; Cuddy et al., 2004). Workplace surveys show that maternal wall bias affects a large proportion of women, with over half of women in some samples reporting that colleagues question their commitment and competence after having children (Williams et al., 2018). Recent data has also found that women in STEMM are three times more likely than men to say they have experienced a decrease in the professional opportunities they are offered after becoming a parent (Torres et al., 2023a).

Maternal wall bias is particularly potent in academic STEMM because it is rooted in the moralized culture of these fields. As previously noted, widely held beliefs about merit in academic STEMM include beliefs that academic science is a vocation that demands and deserves single-minded "work devotion" (Blair-Loy & Cech, 2022). For mothers, far more so than for fathers, childbirth and childrearing are believed to violate their scientific excellence and devotion (Morgan et al., 2021). Among equally productive STEM faculty, mothers are often viewed as less productive (Blair-Loy & Cech, 2022). Research has shown that women often need to work harder to achieve the same recognition as men, and thus "successful women will need to be the most ideal of ideal workers" (Drago et al., 2006). In one study of STEM and non-STEM faculty at a research-intensive university, Kmec (2013) finds that STEM mothers are more likely to say that they must work "very hard" on the job, reporting significantly higher levels of work intensity than STEM fathers as well as non-STEM mothers report, controlling for a variety of relevant factors. Kmec interprets this as mothers in STEM, who are countering a wide set of stereotypes that suggest women and mothers are incongruous in STEM fields, must work harder to avoid perceptions that they do not belong in their jobs (Kmec, 2013). Another study found that women scientists need 64 more impact points than men to be seen as equally competent, amounting to three more publications in *Nature* or *Science* and 20 more publications in less prestigious journals (DesRoches et al., 2010).

Even women who are not mothers may experience maternal wall bias. For instance, in an interview study, highly motivated Ph.D. students without children reported feeling constrained by the stigmatized cultural "specter" of motherhood (Thébaud & Taylor, 2021). Women and men interviewees in the study reported that women graduate students were more likely than their men counterparts to face warnings from faculty advisors that parenthood is incompatible with scientific excellence. For instance, one study participant shared that her faculty advisor, who is a father, told her "There's more to life than babies . . . you should have a passion for science that should be driving you more than . . . family." Another participant in the study shared that her advisor told her "I hope you don't have a kid during grad school" because "[I don't] know how any woman would graduate when they have a kid." Such comments encouraged some women to hide or constrain their parenthood plans or to leave academic STEM (Thébaud & Taylor, 2021).

Though maternal wall bias affects even women whose work patterns do not change after they have children, many women's work patterns do change. That is because academia defines the ideal worker as someone who takes no time off for childbearing or childrearing. The result is work-family conflict, which is exacerbated in academia due to rigid "up-or-out" career tracks, and further exacerbated in STEMM due to the extremely long work hours, including for "night science," that is, experiments that need to be tended even into night hours (Williams, 2000). Yet when mothers respond by attempting to cut back, often they are driven out of academia altogether or onto research tracks that lack the prestige, and often the benefits, available to tenure-track professors (including maternity leave) (Zheng et al., 2022). This corrodes the quality of women scientists' jobs at just the same time when they are struggling to perform as ideal workers without the support for family caregiving available to most men. In fact, the largest leak of women out of the STEM pipeline is when they start families (Goulden et al., 2011). Women Ph.D.s with young children are four times more likely than women without children to leave the labor market entirely (Wolfinger et al., 2009). More recent research found that nearly one-half of new mothers left full-time work in STEMM following the birth of a child, compared with one-quarter of new fathers (Cech & Blair-Loy, 2019).

Given these findings, it is unsurprising that maternal wall bias and motherhood penalties carry economic consequences. Research shows that women often experience sudden, large, and persistent drops in labor market earnings with the arrival of their first child. For example, using 1976–2017 Panel Survey of Income Dynamics, or PSID, data, Cortés and Pan (2020) estimate about 40 percent lower earnings for mothers relative to fathers 5 to 10 years after the arrival of the first child. The actual dollar amounts of these employment-related costs to women were calculated through a recent dynamic microsimulation study of women born between 1981 and 1985 who provide unpaid care to minor children and parents, parents-in-law, and spouses/unmarried partners with care needs (Johnson et al., 2023). These employment-related caregiving costs to women average $295,000 over a lifetime—80 percent of these lifetime employment-related costs are due to lost earnings ($237,000) and 20 percent are due to lost retirement income from Social Security and employment-based plans ($58,000) (Johnson et al., 2023). These employment-related costs are particularly high for mothers with multiple children and for "well-educated mothers, who generally earn higher wages than less-educated ones. Lifetime costs average $420,000 for college-educated mothers, $202,000 for mothers who completed high school but did not attend college, and $122,000 for mothers who did not complete high school" (Johnson et al., 2023). Such motherhood penalties have also been documented among highly educated professionals, such as STEMM professionals (Cech and Blair-Loy, 2019), M.B.A. graduates (Bertrand et al., 2010), and law school graduates (Azmat & Ferrer, 2017).

Women of color may be most affected by these barriers, given studies documenting that they encounter higher levels of bias triggered by both gender and race, compared with White women (Williams, 2014). Racialized conceptions of motherhood as well as distinct histories of caregiving along racial lines shape the ways in which maternal wall bias is experienced by women of different races (Williams et al., 2020). In a study interviewing 60 women of color in science, technology, engineering, and mathematics,

Williams (2014) examined experiences of the maternal wall among Black, Latina, and Asian women. While some similarities to White women emerged (e.g., similarly high rates of White women and women of color reported feeling their career commitment was questioned after giving birth to or adopting a child), important distinctions also arose in interviews. For Black women, for example, higher rates of single motherhood and stereotypes of this status were salient. One respondent pointed out that a man with three children "will be treated like a breadwinner," but a single Black woman with three children will not be treated kindly (Williams, 2014, 203). Latina respondents were affected by close cultural associations of Latinas and motherhood and particular expectations both in the workplace and in their home lives that they would have many children that could reduce their time and commitment for paid labor. They also are more likely to face greater career risk and stigmatization for not being seen as career competent when using flexible and alternative work arrangements, which may further marginalize them from other colleagues for working differently than the norm (Kossek et al., 2023). For Asian women, findings were more complex, as stereotypes of strong familial orientation countered stereotypes of the model minority predicting strong commitment to work (Williams, 2014).

While women of color can face greater challenges associated with maternal wall bias in the workplace, the distinctions between White women and women of color do not always place women of color at a disadvantage. Cultural conceptions of family obligation as well as strong community ties provide a broader community of support for many mothers of color (Williams, 2014). Many immigrant Asian women respondents noted the benefit of their own parents' willingness to come to the United States to help care for grandchildren, and Black women described a much broader circle of caregivers they could turn to for support, including extended family and church communities (Collins, 1991; Stack, 1974). This offers important resources that can provide significant aid to individuals but does not offset or reduce the significant barriers they encounter due to intersecting biases based on race/ethnicity and gender.

Gender, Race/Ethnicity, and Other Caregiving Biases

Caregiving, of course, is not confined solely to care for children. Women also carry a major load in caregiving for adults—whether in the form of spousal care, care for adult dependents, or care for aging parents or other loved ones. Many women also find themselves providing what has been termed "sandwich care," or care for both young children and adult dependents/aging relatives or kin (Pierret, 2006; Suh, 2016). One study of hiring discrimination found that sandwich caregivers were evaluated most negatively by employers in all occupations, regardless of whether those occupations were dominated by men or women (Henle et al., 2020). One randomized control study found that sandwich caregivers who work in health care also caring for older adults are likely to report the highest levels of psychological distress and be in the most need of workplace interventions that are designed to provide more workplace social support for the family role (Kossek et al., 2019). While research has shown an increasing move toward greater sharing of childcare responsibilities by men, a similar move in the direction of gender parity has not been observed in older adult care, including care for a woman's husband's parents (Grigoryeva, 2017). Women are also more likely than men to be the sole caregiver or provide most of the care for an adult family member and to provide a greater intensity of care than men (AARP & National Alliance for Caregiving, 2020). And while the intensity of care men provide is most influenced by the presence of other caregivers, the intensity of care women provide is more often influenced by constraints on their time from paid labor or other caregiving demands (Grigoryeva, 2017).

Norms of care for aging parents and other family members also tend to be weaker among White Americans compared with Americans and immigrants of color, and research has found that caregivers of color report stronger beliefs in filial obligation than do White caregivers (Pinquart & Sörensen, 2005). There is also some evidence that intensity of adult care varies by race/ethnicity. Though estimates vary, a few studies have reported that Black and Hispanic women on average perform higher levels of intensity of adult care compared with White caregivers (Cohen et al., 2019). Black caregivers are also much more likely to provide informal care beyond immediate family members to others such as friends or church members (Cohen et al., 2019; McCann et al., 2000). Together, this can produce differences in the degree to which caregivers are engaged in care of adults by race/ethnicity.

This reality was evidenced in the interviews conducted as part of this study. Broader systems of advantage and disadvantage by race/ethnicity, gender, and class heavily shaped caregivers' ability to manage conflicting career-caregiving demands. Caregivers of color suggested that these systems of privilege simultaneously influenced the structural and interpersonal likelihood that a STEMM scholar would be faced with intensive caregiving conflict as well as the extent of guidance or institutional support they might receive in facing such conflict. One respondent discussed how privilege worked to allow greater space to manage competing demands for some, while leaving others with much less support:

> "The structural issues that exist are, either your privilege gives you the space to be able to meet the standards and expectations that have been institutionally inherited for years—that really only are conducive to a White man who has a wife at home, so it's easiest for them to achieve those—or the people who get the strategic advice and guidance are receiving it because they have some sort of social connection to the people who are in the know."

Caregivers of color relayed how, in the context of longstanding structural racism and exclusion in their institutions and disciplines, it seemed that they were expected to fail when negotiating conflicting career-caregiving demands. This was seen in how departments and institutions responded to challenges caregivers of color faced as well as assumptions made about how they were doing:

> "From a more, like, systemic discrimination aspect, I felt [that] my institution was just like kind of waiting for me to fail ... I was juggling a lot of things, and they were just waiting for me to, like, drop everything ... they highlighted the times that I didn't meet expectations a lot more than, like, all the other times that I did, or that I did publish, or that, you know, I did do extremely well ... it was just interesting how often they were quick to say like, "Oh, it's because she's got kids." They're like, "Oh, it's because she had kids during the program," or it's "Oh, it's because her dad's, you know, sick." So, I felt like ... they used it against me."

Caregivers of color also repeatedly highlighted the central importance of robust institutional support for caregivers as a core equity issue. For institutions aiming to build belonging, equity, and inclusion, they argued recognizing the cultural value of caregiving and supporting scholars in providing that care was critical. As one respondent noted:

> "Many cultures like my own—Hispanic, I believe African Americans, Native American—we are very committed to family. And many of us are now taking care of somebody in the family, especially when we start getting older.... An incredible recruitment

tool to increase diversity and a sense of belonging to people from different cultures [is] health insurance or packages that will cover extended family members, that kind of support.... With something like that, I think institutions will be able to recruit the population of faculty and staff that they want to have to increase diversity. Not having [these] kinds of policies has been very difficult."

Flexibility Stigma

Those in academic STEMM who have taken advantage of family supportive policies sometimes face penalties in the form of a lack of support from department chairs and colleagues, as well as retaliation in career assessments (Kachchaf et al., 2015; Kossek & Lee, 2022; Sallee, 2012; Williams, 2005, 2014). Cech and Blair-Loy (2014) describe such penalties for those who take care-related workplace accommodations as "flexibility stigma" (see also Blair-Loy and Cech, 2022; Williams et al., 2016). They find in their research that individuals in academic STEMM who use informal or formal arrangements to balance care for their children are seen by colleagues as less committed to their careers. Williams et al. (2016) argue that resistance to flexible work policies is "fueled by identity threat"; that is, work is often so intimately tied to identity that efforts to reform the hegemonic culture of (over)work can be threatening to those who developed their identities around the original model (of the ideal worker).

As mentioned throughout this chapter, though this stigma has ramifications regardless of gender, women face unique and disproportionate burdens. Research has shown that women with children are often the "default parent" even in two-parent households where both parents are employed outside the home (Calarco et al., 2021; Hosek & Harrigan, 2023; Rinaldo & Whalen, 2023). This default status increases mothers' need for flexibility as they are seen as the go-to parent for unexpected events such as illness, school closures, or other disruptions. Given that mothers are more likely to need this kind of last-minute flexibility, this opens them up more frequently to the potential for flexibility stigma.

Despite the existence and usage of leave and caregiving policies, STEMM men and women caregivers who experience this stigma are less likely to be satisfied at the institution, less likely to feel work-life balance, less likely to plan to stay at the institution, and more likely to consider leaving academia for industry (Cech & Blair-Loy, 2014). At the most extreme, some in academic STEMM may face family responsibilities discrimination based on their family caregiver status (WorkLife Law, 2017). For example, Williams and Norton (2010) describe one lawsuit after a woman faculty member took parental leave and delayed her tenure clock:

> "Despite unanimous recommendation from her tenure committee and endorsement from the dean, she was refused tenure upon her return. The provost ... allegedly told another professor that the mother's decision to "stop the clock" was a "red flag," and the department chair wrote in a memo that [the faculty member] "knew as the mother of two infants, she had responsibilities that were incompatible with those of a full-time academician."

Institutions that do not address this stigma are more likely to lose valuable STEMM professionals, and those who do may be advantaged in attracting them.

THE IMPACT OF CULTURAL SCHEMAS ON THE ORGANIZATION OF STEMM

Cultural schemas shape the organization of STEMM work and learning environments in ways that reflect assumptions about who is carrying out the work and how their lives are organized. Since ideal worker norms and expectations of devotion in academic STEMM implicitly assume a lack of outside obligations, such as that for caregiving, expectations for career trajectories and everyday work and educational arrangements often follow this assumption. So far, this report has largely focused on the influence of cultural norms, but the solutions are in acknowledging they exist and taking action to limit their consequences.

Cultural effects show up as formal and informal policies in career tracks, grant eligibility and timelines, and work emphasis. Today's academic careers are organized in a neat, linear fashion—undergraduate education, graduate school, postdoctoral work, and entering the tenure track with a linear ascension from assistant to associate to full professor—that does not recognize how care responsibilities may impede progress (Winslow & Davis, 2016). Indeed, the time-delineated tenure clock often hits academics precisely at the life stage when many would ideally have children (often having delayed this during training) (Beckerle et al., 2011; Jacobs & Winslow, 2004; Mason et al., 2019; Winslow & Davis, 2016). Academic promotion and recognition often have milestones that are difficult to meet as a part-time academic, and tenure clocks generally do not tolerate periods of lower career intensity. Addressing these structural barriers to family caregivers in STEMM requires that higher education institutions, federal agencies, and accrediting bodies acknowledge they exist and initiate appropriate countermeasures.

The timing of grants often fits assumptions about outside obligations as well and proves challenging for caregivers. Many early-career grants place time-delimited requirements on when a person is eligible to apply for funding. Such limitations often overlap with timelines for when an individual may be seeking to start or building a family (Kossek & Lee, 2020). Even for grants that are not limited to early-career researchers, funding and deadlines can also be challenging for faculty with caregiving responsibilities. In acknowledging this structural barrier, some federal granting agencies and individual foundations have been taking action and revising policies to ensure that academic STEMM faculty with caregiving needs are able to take leave and extend their grants and eligibility for funding (see Chapter 4 for specific examples of these policies). Additionally, some federal agencies provide funding for childcare or replacement of faculty or staff on care leaves (Beckerle et al., 2011). Unfortunately, not all institutions make it easy for researchers to make use of this flexibility and without it, meeting grant deadlines can be a challenge.

Additionally, even informal policies and practices can reflect assumptions that disadvantage faculty, staff, and students with caregiving responsibilities (Ecklund et al., 2012; Sallee, 2012). Informal policies such as scheduling talks and meetings early or late in the day, expecting or encouraging time in the office or laboratory on the weekends or late at night, or policies limiting access to flexibility or asynchronous work or learning also may reflect and contribute to a culture that does not acknowledge caregiving responsibilities (Vos et al., 2021). Consideration of these care demands in work scheduling expectations and greater flexibility would ensure broader participation in these core research activities by faculty with care demands (Kossek & Lee, 2020, 2022).

STEMM faculty, graduate students, and postdocs and other trainees with caregiving responsibilities also find it difficult to attend conferences and engage in travel and networking that is often key to academic careers (Beckerle et al., 2011; Calisi & Working Group of Mothers in Science, 2018; Fuentes-Afflick et al., 2022; Lubitow & Zippel, 2014; Tower & Latimer, 2016; Winslow & Davis, 2016; Xu & Martin, 2011; Zippel, 2017). In another study, at least half of the full-time faculty with children at a large research university noted that lack of childcare meant that they were unable to plan or attend research-related travel, submit to a conference, or accept an invitation to give a talk; this effect was stronger for those faculty who were partnered with another academic (Tower & Latimer, 2016).

IMPACTS OF COVID-19: EXACERBATING LONGSTANDING INEQUITIES IN THE PROVISION OF CARE

Although caregivers in academia have always faced substantial challenges, the COVID-19 pandemic intensified and put a spotlight on these challenges with the nationwide disruption of schooling, childcare, and older adult care facilities (National Academies of Sciences, Engineering, and Medicine, 2021). Early in the pandemic, researchers identified that while everyone was affected by the pandemic, certain groups were experiencing a disproportionate amount of pressure. Research showed that women, in particular, faced substantial work disruptions due to caregiving (National Academies of Sciences, Engineering, and Medicine, 2021). At the same time, people of color, who generally experienced higher rates of COVID-19 and COVID-19-related deaths, were more likely to experience intensified caregiving for family members suffering from COVID-19, as well as to offer bereavement care (Aburto et al., 2022; Douglas et al., 2022).

In STEMM, the impact of the COVID-19 pandemic threatened the hard-won progress toward greater gender equity in these fields. To quote a National Academies (2021) study on the impact of the COVID-19 pandemic on women's research careers:

> "The evidence available at the end of 2020 suggests that the disruptions caused by the COVID-19 pandemic endangered the engagement, experience, and retention of women in academic STEMM, and may roll back some of the achievement gains made by women in the academy to date."

Several studies have documented the disproportionate effect of the pandemic on caregivers, particularly women, in STEMM. For example, one study of papers published in medical journals in 2019 and 2020, which compared authorship for papers about COVID-19 with papers in the same journal the previous year, showed a significant decrease in women's engagement as first authors (Andersen et al., 2020). Similarly, a survey of 1,185 medical, graduate, and health professions schools at one university showed that women were more likely than men to consider going part-time or leaving employment during the pandemic, with the strongest effect among women with children (Matulevicius et al., 2021). Both men and women with children were also more likely to consider going part-time or leave employment, and working parents as well as women were more likely to turn down leadership opportunities (Matulevicius et al., 2021). As Matulevicius and colleagues note (2021, 6):

> "This association of both gender and parenting with increased perceived stress may disproportionately decrease the long-term retention and promotion of junior and mid-career women faculty."

Another fall 2020 survey of faculty at a large urban U.S. university system that includes 2- and 4-year institutions included 3,219 participants (Skinner et al., 2021). While all groups noted spending less time on research during fall 2020 than they expected, there were important gaps in research products by gender and caregiver status (Skinner et al., 2021). Early-career scholars were also more likely to note that their career options were being negatively affected by the pandemic (National Academies of Sciences, Engineering, and Medicine, 2021).

One of the most critical challenges during the pandemic was the increasingly "boundaryless" nature of work (National Academies of Sciences, Engineering, and Medicine, 2021). While everyone in academic STEMM experienced a loss of boundaries between work and home, this challenge was particularly exacerbated for caregiving women who remained primary caregivers in many households (National Academies of Sciences, Engineering, and Medicine, 2021). These problems were further exacerbated for mothers of children with disabilities, since in many instances the move to online classrooms could not fully meet the educational and developmental needs of these students (Schneider et al., 2021). These challenges led to a variety of negative outcomes regarding increased workload, decreased efficiency, and negative effects on personal well-being (National Academies of Sciences, Engineering, and Medicine, 2021). Further, a national study of academic STEMM women found that when women faculty experienced disrupted boundaries and gender inequalities in the division of labor on and off the job, the result was a greater likelihood they felt the strain of not having enough support for caregiving. In turn, this led them to withdraw from their jobs, experience burnout, and contemplate leaving their occupation (Kossek, Perrigino, & Rock, 2021).

On April 10, 2023, President Biden signed a bill that ended the COVID-19 national emergency. But despite the pandemic's official end, it remains crucial to consider the lessons learned from this period. The disproportionate impact of the COVID-19 pandemic on caregivers, and especially women caregivers, demonstrated how ill prepared STEMM educational and work environments are to handle such a disruption. The pandemic has highlighted the need for comprehensive policies and practices in support of caregivers in STEMM and made more visible the longstanding challenges in STEMM fields faced by caregiving students and professionals. It is critical that attention to these remain priorities in a postpandemic world. Moreover, reports indicate that the lingering effects of the pandemic remain; a member of the National Science Board observed that the pandemic may have set back U.S. women's science careers permanently (Richmond, 2020).

BALANCING THE BENEFITS AND DRAWBACKS OF IDEAL WORKER NORMS

Discussions of the challenges associated with current norms perhaps inevitably produce questions about whether the status quo has important benefits that we might lose if it were to change. As discussed throughout this chapter, there are substantial costs to the ideal worker status quo, in terms of long-term productivity, creativity, and health of those in academic STEMM, and on whether academic STEMM ends up selecting and promoting those with the greatest talent or those who can work a certain schedule. Some may argue, however, that ideal worker norms increase productivity because of returns to experience and long hours and, in this way, benefit the scientific enterprise by increasing scientific output.

Certainly, any set of norms comes with both potential benefits and potential costs. The committee, however, is not arguing that the appropriate response is to replace rigid norms of overwork with infinite flexibility. Instead, the key question is how to create academic STEMM careers that can still capture the benefits of intense focus on work while avoiding the documented costs in health, quality of work, and the exclusion of certain groups from the workforce in part through recognition of the value and importance of unpaid labor. Finding the optimal balance can include reconsidering the way metrics such as publications and citations are evaluated, thinking carefully about the timing of meetings and other deadlines, introducing policies that protect promotion opportunities for those who need to take leave such as stop-the-clock, and accepting the reality that all scientists—as human beings—have predictable and unpredictable needs to care and be cared for during certain periods of their lives. The committee discusses these and other policies and practices that support family caregivers in the next several chapters.

SUMMARY OF FINDINGS FROM CHAPTER 3

1. Together, we see that the broader cultural context in STEMM, which expects an ideal worker who can devote their full time and attention to work, with little to no outside demands, creates challenges for caregivers of all genders, both in terms of navigating time constraints and of facing potential bias and discrimination because of their caregiver status in the workplace. These cultural norms also set the stage for the structure and policies of academic STEMM in ways that can create greater challenges for family caregivers. And coupled with particularly intense expectations of devotion within academic STEMM, they additionally interact in significant ways with the uneven burden of family caregiving responsibilities by gender and race/ethnicity. All of this is further compounded by the experiences of the COVID-19 pandemic, with disproportionate impacts on caregivers, particularly women, women of color, and communities of color. The result is that those who are already most marginalized within academic STEMM fields are further marginalized for providing necessary care for family members and loved ones. Ideal worker norms in STEMM are characterized by an assumption that students and academicians will show devotion through working long hours, being constantly available and visible, and pursuing unbroken career trajectories. These norms shape cultural conceptions of how people should approach working and learning in STEMM and perpetuate rigid structures for career trajectories, daily work schedules, and research timelines in ways that amplify tensions with expectations of caregiving.
2. The expectations and structure of today's academic STEMM workplace are harmful to all groups, but they have different consequences by gender. For women, they fuel assumptions that women are not ideal workers because they have caregiving responsibilities, and therefore are not qualified for professorial positions or for promotions. For men, they fuel pressures to work the same kind of hours and patterns as someone who has no ongoing responsibilities for daily care of others.

3. Maternal wall bias is particularly potent in academic STEMM because it is rooted in the culture of these fields. Widely held beliefs about merit in academic STEMM include beliefs that academic science is a vocation that demands and deserves single-minded "work devotion." For mothers, far more so than for fathers, childbirth and childrearing are believed to violate their scientific excellence and devotion. The result is real and substantial financial loss, premature academic career pruning for women (e.g., choosing non-tenure-track positions, moving to industry), and a true voltage drop for STEMM academic productivity.
4. Women of color may be most affected by maternal bias. Racialized conceptions of motherhood, as well as distinct histories of caregiving along racial lines, shape the ways in which maternal wall bias is experienced by women of different races and ethnicities.
5. The COVID-19 pandemic intensified caregiving versus career challenges, including disruptions in schooling and childcare and in caring for sick relatives. Research shows that women faced substantial work disruptions due to caregiving, and faculty of color were more likely to experience intense caregiving for family members suffering from COVID-19, as well as loss. It demonstrated that current support for academic STEMM professionals with caregiving responsibilities was already broken.

4

Current Laws, Policies, and Practices to Support Family Caregivers

This chapter draws substantially from the research paper "Comprehensive Literature Review of Current and Promising Practices to Support Unpaid Caregivers in Science, Technology, Engineering, Mathematics, and Medical STEMM," by Jessica Lee, J.D., Erin Frawley, M.Ed., and Sarah Stoller, Ph.D., which was commissioned for this study.[1]

The current landscape of laws and policies that support family caregivers consists of a variety of piecemeal efforts that provide a degree of support for caregivers, but are neither comprehensive nor coordinated, resulting in significant gaps. These efforts exist at many different levels, from broad national laws to policies adopted by outside agencies, funders, and accrediting institutions that influence universities to individual universities themselves. This chapter provides an overview of the policies and practices that support family caregivers. It starts by describing the external forces that can shape and influence university policies, including federal and state laws, the policies of federal agencies and other funders, and accrediting bodies. It then turns to universities themselves, acknowledging that there is no one approach to work-family policies at U.S. institutions, but rather an array of different elements that can be implemented in a variety of ways. The goal of this chapter is to provide a clear understanding of current laws and policies by outlining the existing status quo for caregiving support. Chapters 6 and 7 build on this information to detail best practices and innovative ideas, respectively.

CURRENT FEDERAL AND STATE LAWS

In the United States, no single law establishes and defines the full set of rights and protections of family caregivers. Instead, a set of mandates and

---

[1] The full paper is available at https://nap.nationalacademies.org/resource/27416.

regulations at the federal, state, and local levels provide legal protections and accommodations for family caregivers. These mandates and regulations are not coordinated nor are they connected, resulting in a complex patchwork that creates challenges for individuals seeking to understand what protections apply to them. This chapter summarizes key laws related to caregiving, including protections for university employees (see Table 4-1) and protections for students (see Table 4-2). Additional information is available in Appendix A.

TABLE 4-1 Federal and State Legal Protections for Employees

Type of Law	Federal/State	Key Provisions
Childbearing and Caregiving Leave	At the federal level, caregiving leave protections mainly fall under **Title VII and the Family and Medical Leave Act (FMLA),** but may also be covered under **Title IX, the Pregnant Workers Fairness Act,** and the **Americans with Disabilities Act** for pregnant individuals At least 16 states provide job-protected leave for caregiving employees	**Title VII of the Civil Rights Act of 1964** requires paid childbearing leave for all birth-parent employees as long as deemed medically necessary by their health care provider if an institution grants disability leave for conditions other than childbearing. This typically translates to 6 to 8 weeks of paid leave and covers employers with 15+ employees The **Family and Medical Leave Act** requires covered employers to provide eligible employees with up to 12 weeks of unpaid, job-protected leave in a 12-month period. At institutions with 50+ employees, all full-time and some part-time faculty are covered **Title IX** requires educational institutions to provide employees with leave for pregnancy or related conditions The **Pregnant Workers Fairness Act** requires reasonable accommodations, including leave, for those affected by pregnancy and related conditions The **Americans with Disabilities Act** requires accommodations, including leave, for those with a disability, including pregnancy-related disabilities

TABLE 4-1 Continued

Type of Law	Federal/State	Key Provisions
		State family and medical leave laws vary and are typically similar to the FMLA in job protection, but provide for paid leave as well as expanded eligibility by reducing employer size thresholds or the length of time an employee must have worked to be eligible for leave
Maternity Accommodations	At the federal level, maternity accommodations fall under the **Pregnant Workers Fairness Act** and the **PUMP for Nursing Mothers Act**	The **Pregnant Workers Fairness Act** requires employers to grant accommodations to employees affected by pregnancy and related conditions, such as infertility, miscarriage, pregnancy loss and abortion, childbirth and recovery, postpartum depression, and lactation. It covers employers with 15+ employees
		The **PUMP Act** is an amendment to the Fair Labor Standards Act that requires employers of all sizes to provide employees with lactation breaks as needed and a lactation space that is not a bathroom and is free from view and intrusion
Antidiscrimination Protections	At the federal level, **Title VII** and the **Americans with Disabilities Act** provide key discrimination protections	**Title VII** prohibits employers from discriminating on the basis of sex, race, color, national origin, or religion. Case law establishes that it covers discrimination against mothers, and can cover discrimination against other caregivers
	At the state and local levels, over 200 jurisdictions have laws prohibiting discrimination based on caregiver status or family responsibilities	The **Americans with Disabilities Act** prohibits discrimination based on an employee's association with an individual with a disability, which includes caregivers discriminated against because they are caring for a disabled child, partner, or other individual with a disability. This covers all employers with 15+ employees
		In 252 American jurisdictions covering roughly 1 in 3 workers, state and/or local laws prohibit discrimination against caregivers. The specifics of these laws vary widely

TABLE 4-2 Federal and State Legal Protections for Students

Type of Law	Federal/State	Key Provisions
Caregiving Leave	At the federal level, caregiving leave is provided under **Title IX**	**Title IX** of the Education Amendments of 1972 prohibits discrimination on the basis of sex and requires educational institutions to provide their students and trainees with childbearing leave for as long as medically necessary
Maternity Accommodations	At the federal level, maternity accommodations are provided under **Title IX**	**Title IX** requires educational institutions to provide their students and nonemployee trainees with accommodations and academic adjustments when needed due to pregnancy and related conditions
Antidiscrimination Protections	At the federal level, antidiscrimination protections are provided under **Title IX**	**Title IX** prohibits discrimination and harassment on the basis of sex, which includes discriminatory treatment of students based on pregnancy

As summarized in Tables 4-1 and 4-2, the United States has a system of federal, state, and local protections that can be relatively comprehensive but piecemeal and inconsistent across locations. Multiple laws provide employees with leave, accommodations, and protection against discrimination; state and local laws provide additional protections. Faculty, staff, and other university employees in certain states have greater protections for caregiving responsibilities than those in others. For example, 12 states and Washington, D.C., have a law requiring paid leave for new parents and family caregivers, though even in these states, laws typically have caps on benefit amounts and are generally unable to replace a faculty member's pay (A Better Balance, 2023). In contrast, caregiver protections for students fall largely under a single federal law, Title IX of the Education Amendments of 1972 (Mason & Younger, 2014). While Title IX affords a range of important protections, it is not comprehensive. For example, student leave under Title IX is provided and protected for pregnancy and related conditions, but Title IX does not provide the right to leave for other caregiving responsibilities, such as caring for a parent, a sibling, or an adult child.

Together, various laws and regulations provide some degree of support and protections for family caregivers and provide a set of minimum requirements that colleges and universities must meet and adhere to as they build their own practices. Still, current laws are a patchwork that can be challenging to navigate, for faculty and university officials as well as employees and students. And current laws are incomplete, especially regarding coverage of caregiving responsibilities other than for children and regarding paid family and medical leave. Title IX does not consider caregiving responsibilities not related to pregnancy and parenting, and similarly, many laws regarding accommodations for both students and employees are maternity related, which does not consider accommodations people may require for other caregiving responsibilities. Additionally, the United States is the only Organisation for Economic Co-operation and Development (OECD) country that does not provide any form of federal paid family and medical leave. In contrast, on average across the OECD,[2] mothers are entitled to nearly 51 weeks of total leave time, with an average of 19 weeks of paid maternity leave and 32 weeks of paid parental and home care leave.[3] Fathers or non-birthing parents are entitled to around 10 weeks total on average, composed of an average 2 weeks of paid paternity leave and 8 weeks of paid parental and home care leave (Organisation for Economic Co-operation and Development, 2022). Along with parental leave policies, more than half of OECD countries provide paid leave for the care of sick children or other family members, though the amount varies considerably (Organisation for Economic Co-operation and Development, 2020).

FEDERAL AGENCIES' AND OTHER FUNDERS' POLICIES SUPPORTING CAREGIVERS

Along with federal and state regulations, caregivers in academic science, technology, engineering, mathematics, and medicine (STEMM) may find support in the policies of federal funding agencies (e.g., National Institutes of Health [NIH], National Aeronautics and Space Administration [NASA], National Science Foundation [NSF], National Institute of Standards and Technology) as well as private funders and foundations (e.g., the Doris Duke Charitable Foundation, Henry Luce Foundation, American Cancer Society, Afred P. Sloan Foundation). For those receiving grants or fellowship money to support their work, federal agencies and other funders may implement a variety of policies to ensure flexibility for grant recipients with

---

[2] These averages include the United States, which offers 0 paid weeks across all categories.

[3] *Parental leave* refers to a leave of absence for employed parents that is supplemental to maternity and paternity leave. *Home care leave* refers to leave generally following parental leave to allow at least one parent to stay home and provide care to a young child under 2 or 3 years of age.

caregiving responsibilities as well as provide opportunities to use funding for purposes related to childcare.

Federal agencies and other private funders provide important resources in support of research, and this funding often supports the salaries of principal investigators, staff, trainees, and students. While it is earmarked for a particular project, individual, or group, some funders stipulate flexibility in use of funds for caregiving needs or provide access to additional resources to meet these needs. For example, the NIH (2021) has several caregiver-friendly initiatives, which include the following:

- Reimbursement of caregiving-related costs
- Additional funding awards to support caregiver leave
- Funds to offset the cost of childcare
- Reentry programs to support grant recipients who have taken time away for caregiving
- Extensions on timelines for early-stage investigator award eligibility

NIH has also recently undergone policy changes in response to pushback from postdoctoral fellows in a letter urging better family supportive policies to prevent women from leaving the academic workforce (Guo et al., 2023). This letter detailed (1) the lack of paid parental leave, (2) the lack of support during the transition back to work, and (3) the high cost of childcare as barriers to remaining at work. Accordingly, the NIH has implemented several reforms over the past 2 years. In particular, National Research Service Award postdoctoral fellows (F30, F31, F32) may now apply for up to $2,500 per budget period to defray the costs of childcare (National Institutes of Health, 2021). Notably, this amount would be expected to cover approximately 2 months of infant care at a childcare center (Child Care Aware of America, 2022).

Similar policies are also in place among other federal agencies. The NSF, for example, has a Career-Life Balance Initiative instituted in 2012 that has worked to organize and disseminate information on policies that provide flexibility for those caring for dependents. Through this initiative, individuals with family caregiving responsibilities are eligible to take leaves of absence, access no-cost extensions, or delay starts to grants, and can apply for supplements up to $30,000 to hire temporary support for those on leave (National Academies of Sciences, Engineering, and Medicine, 2023; National Science Foundation, 2015).

NASA similarly provides flexibility in support for grant recipients with family caregiving responsibilities. All grant recipients can make use of one no-cost extension without requiring approval, though subsequent extensions do require approval. Though NASA does not have a centralized source of funding to provide supplements to support bringing on additional personnel for those on leave, they can consider these case by case. Other policies such as the use of award money for dependent care or to pay for leave are allowable only when a recipient's host institution allows this (National Aeronautics and Space Administration, 2021).

Private funders have targeted grants for investigators with family caregiving responsibilities, and some have created programs specifically designed to support individuals with substantial caregiving responsibilities. The Doris Duke Charitable Foundation, for example, created its Fund to Retain Clinical Scientists (FRCS) in 2015. This fund provides grants to medical school recipients to establish an FRCS program that provides supplemental research support to early-career scientists with family caregiving responsibilities (Myers, 2018). The funds can be used to hire technicians and research coordinators and to buy back clinical obligations to provide more time for research (Jagsi et al., 2018, 2022).

While some federal and private funders provide easily accessible information dedicated to family-friendly initiatives,[4] such as on well-organized web pages devoted to these policies, many funders only include information about policies aimed at caregivers within the pages of long handbooks on grant requirements, if this information can be found on their websites at all.

ACCREDITATION AND CERTIFICATION BOARDS

Accreditation and certification boards also play a role in shaping university policies. These boards set requirements that a university or program must meet to receive accreditation or certification. In recent years, key accrediting bodies have adopted policies that affect caregivers in medicine. For example, in 2022, the Accreditation Council for Graduate Medical Education (ACGME) began to require all accredited training programs to offer 6 weeks of paid leave to residents and fellows for parental and caregiving leave. This policy was essential given the heterogeneity and inadequacy of leave typically offered to resident physicians in training (Magudia et al., 2018). It allows medical residents to be able to afford the leave that they need (Ortiz Worthington et

---

[4] For example, the Office for Research on Women's Health at the NIH provides a web page detailing the range of policies in support of caregivers. See https://grants.nih.gov/grants/policy/nih-family-friendly-initiative.htm.

al., 2019), and it reduces gender disparities in the process, as research shows that men are less likely to take leave when it is unpaid (Halverson, 2003).

The ACGME policy complemented an initiative of the American Board of Medical Specialties (ABMS) that took effect in 2021, requiring its member boards to develop written policies stating the time required for physician trainees to become eligible for board certification. That initiative further mandated that these eligibility requirements permit a minimum of 6 weeks of time away from training once during training for member boards with training programs of at least 2 years, for the purposes of parental and caregiver leave, without exhausting other forms of time off (e.g., vacation and sick leave) and without extending training (American Board of Medical Specialties, 2021). Prior to this, leave for resident physicians was often inaccessible, in practice, since taking time away from training could create cascading challenges that extended training time even further (Jagsi et al., 2007). The ABMS policy further supports the continued education of residents who take leave by encouraging the scheduling of subspeciality fellowships after July to allow those who must extend their training to have access to the fellowships.

INSTITUTIONAL POLICIES

Individual academic institutions often offer policies, programs, and practices intended to support employees, students, postdocs, and trainees with caregiving responsibilities. While compliance with federal and state laws and with accrediting bodies can drive some degree of convergence in the policies offered by various institutions, policies can still vary greatly between institutions and there is no universal approach to supporting family caregivers. The section below outlines the main types of policies commonly seen at academic institutions and the ways in which these policies may be structured. In general, policies can be thought of in four main categories: (1) policies providing caregiving leave, (2) policies providing accommodations and adjustments to regular responsibilities and timelines for caregivers, (3) policies providing direct care support, and (4) policies that aim to prevent or respond to discrimination and harassment based on a person's caregiver status.

Policies Related to Leave

Caregiving Leave Policies

Today there is no single approach to caregiving leave across colleges and universities. Leave lengths, pay rates, and eligibility vary significantly institution by institution (Riano et al., 2018), and policies can even vary within the same institution (Anthony, 2011). It is also not clear exactly how many U.S. academic institutions offer leave policies of any kind, and existing data have largely focused on a narrow set of institution types or departments. Additionally, existing data largely speak to parental leave policies following the birth of a child rather than broader caregiving leave. These data, however, provide some indication that a substantial proportion of schools may not offer leave, a significant gap that leaves many caregivers underserved and indicates a lack of compliance with federal policy in some cases. For example, a 2019 study of the top 25 schools of public health in the United States found that 80 percent had paid childbearing leave for faculty, but only 48 percent provided this leave for staff (Morain et al., 2019).[5] These institutions averaged a leave term of 8 weeks, just slightly more than half of the 14 weeks paid leave recommended by the American Public Health Association. Further, although these schools had published policies, they were often unclear and difficult to understand (Morain et al., 2019).

Additionally, while most universities now have paid parental leave for faculty, this does not universally extend to postdoctoral fellows and graduate students in STEMM who may simultaneously hold positions as both students/trainees and employees of the university (Lee et al., 2017). There are significant gaps in our knowledge of how universities handle leave for students at the undergraduate and graduate level, but available data suggest that policies offering leave for students are rare. For graduate students, Mason and colleagues (2007) found that only 26 percent of universities in the United States had graduate student maternity leave policies in 2007. As with faculty data, more findings exist looking at specific departments. A 2008 study of sociology Ph.D. programs found that few official policies existed to support graduate student parents (Springer et al., 2009), while a separate examination of sociology programs noted that graduate students typically perceived leave policies as being designed only for faculty (Kennelly & Spalter-Roth, 2006). Given many graduate students do not qualify for the Family and Medical Leave Act (FMLA) due to their student status, in many instances taking a leave of absence may be the only option available (Springer et al., 2009). Additionally, the committee could not find national data on the prevalence of parental and family leave policies at the

---

[5] This distinction between availability of paid leave for faculty compared with staff aligns with general trends seen in leave across the workforce in which those with greater income are more likely to receive paid leave benefits than those with lower income (Klerman, Daly, & Pozniak, 2012).

undergraduate level. Given that institutions often do not have a continuous registration policy, the committee speculated that some institutions may expect students to simply fail to register when leave is needed, since maternity leaves are commonly addressed under standard medical withdrawal policies. Ultimately, there are strong indications that universities largely do not have robust policies in these areas and much more to learn and develop to better support graduate and undergraduate students who need leave.

Family leave tailored to the unique needs of medical students has been explored based on information on school websites. In a recent review of websites, Kraus and colleagues (2021) found that 33 percent listed some form of parental leave policy related to pregnancy, birth, and family at 199 medical schools granting M.D. (doctor of medicine) and D.O. (doctor of osteopathic medicine) degrees. Roselin and colleagues (2022) similarly reviewed the websites of 59 highly ranked allopathic medicine schools and found that 46 percent listed leave policies that mentioned "parental needs," while only 14 percent referenced "parental and family leave policies."

Another key gap in leave policies is the difference in access to and use of leave among men and women. Based on one analysis of the experiences of 741 postdoctoral scholars at 63 institutions across the United States, while many postdoc mothers lacked access to paid leave, fathers were at times left out of leave policies (Lee et al., 2017). The report also detailed the pressures fathers faced against taking leave and stereotypes about men and caregiving. In interviews conducted as part of the study, postdocs spoke to experiences of bias in the form of negative comments about taking leave or continued scheduling of meetings during leave periods (Lee et al., 2017). Among faculty, gender-neutral policies have become more common, but at the same time, research has shown the significant role of campus culture in discouraging use of these policies by men (Lundquist et al., 2012). This gender difference is especially consequential given evidence about the effect of parental leave for fathers on gender equality (Gonzalez & Zoabi, 2021; Kotsadam & Finseraas, 2011).

The committee notes that another challenging aspect of drafting a formal leave policy in an academic setting is establishing a leave length that matches the needs of the employee and institution. There is widespread consensus that a leave term of at least 12 weeks is beneficial for employees and their families (this is in alignment with the FMLA's requirement to provide leave for 12 weeks annually for welcoming a new child or caring for a seriously ill family member (U.S. Department of Labor, 1993). But 12 weeks falls just short of a typical university semester, leading institutions to offer leave in semester increments—which works so long as the pregnancies or family illnesses are well timed to the start and end of a semester, which is unlikely. Some universities have allowed employees to donate their sick time to colleagues who have exhausted their allotted leave benefits, such as a program at the University of Alabama at Birmingham (2018). It is not clear how common these programs are, and they still may not be sufficient to ensure faculty can take the full leave they require.

Policies Related to Adjustments and Accommodations

Stop-the-Clock Policies

Concerns about securing promotion and tenure can be a major barrier to academics taking the family leave they need. In a study of more than 1,300 faculty, 33 percent of women opted not to take their designated maternity leave because of concerns about tenure (Koppes Bryan & Wilson, 2015). Tenure clock extension policies have been used to allay concerns about having children and taking leave to support caregiving for children and adults in the pre-tenure years, but controversy persists regarding their efficacy and best practices.

Stop-the-clock (STC), or tenure extension, policies have been among the most widespread institutional interventions implemented to support caregivers in higher education. The first STC policy was introduced at Stanford University in the early 1970s for female faculty members who had babies prior to receiving tenure (Manchester et al., 2013). The aim of the policy was and remains to prevent penalizing birthing and caregiving faculty in the tenure process by accounting for time lost to leave periods and caregiving responsibilities. As of 2005, 86 percent of research institutions offered STC policies (Hollenshead et al., 2005). They range from policies that mainly apply to new parents—typically on a gender-neutral basis—to wider-ranging policies covering circumstances including the birth or adoption of a child, caring for a sick relative, personal illness, and other unforeseen research delays (e.g., Institutional Review Board delays) (Manchester et al., 2013).

While STC policies are now part of a suite of family-friendly benefits offered by a sizable majority of research universities, there is continued debate in the literature surrounding their effectiveness and impact. Most significantly, research has highlighted important implications for promotion. A recent study reported that women and men who used STC were equally disadvantaged in time to tenure; however, being a woman and having taken a tenure extension negatively affected women's promotion to full professor more so than men's, even years later (Fox & Gaughan, 2021). And a 2018 study of economics departments found that STC policies increased the likelihood of men receiving tenure while decreasing women's likelihood at their first academic institution (Antecol et al., 2018). In contrast, a 2013 study at a single university found higher rates of promotion among faculty who used STC policies, but also noted salary losses for these faculty (Manchester et al., 2013). And a 2022 study found that universal and opt-in stop-the-clock policies predict higher proportions of women and particularly women of color among tenured faculty (Gonsalves et al., 2022).

Overall, the findings on STC are mixed but do suggest there could be negative effects of STC policies depending on how they are structured. Importantly, as well, it is not clear what the true alternative should be when evaluating the effects of STC policies. STC policies may slow down tenure processes for those who use them compared with those who did not need or utilize them. However, in the absence of STC policies, faculty who would have used them may have instead left academia to manage their caregiving responsibilities. In Chapter 6, the committee discusses best practices aimed at addressing some of the potential negative consequences of STC policies.

Accommodations, Adjustments, and Duty Modifications

Caregivers in academic STEMM may need changes to how, when, and where work is performed to provide care to others. These changes can be captured by various terms, including *reasonable accommodations*, *academic adjustments*, *active service modified duties*, and *workplace flexibility* depending on who is accessing them and how they are structured.

Caregiving students who desire to take less leave, or who need ongoing changes to be able to stay enrolled while meeting their educational goals, may be entitled to family-responsive academic adjustments. Reducing course load is an example of a common academic adjustment for caregiving students. This could entail extending a student's time to degree and prorating their stipend to match their reduced workload (Springer et al., 2009). At the undergraduate level, student parents often attend part-time to meet the demands of caregiving and work, especially while enrolled at community college (Huerta et al., 2022). For pregnant students, universities may offer accommodations to allow them to participate in classes virtually as well as required accommodations for lactation (The Pregnant Scholar,

2023). Workload flexibility for students is another space where student caregivers may require accommodations. Anecdotal evidence suggests that most institutions do not have formalized policies for allowing students to attend to family needs. Typically, attendance policies are managed at the level of individual faculty advisors.

For faculty, the concept of modified duties was designed to create flexibility in faculty members' workloads by changing job responsibilities without any changes in pay. The most common modification is a reduction in teaching with the expectation that time devoted to the classroom will be reassigned to other responsibilities that allow for more flexible scheduling. Some universities reduce the work assignment for a set period. Others may offer modification on an open-ended basis. One study found that 18 percent of 255 institutions surveyed had implemented a formal modified duties policy (Koppes Bryan & Wilson, 2015). A shared laboratory with two principal investigators can be highly productive while allowing for flexibility (Oldach, 2022). (See Chapter 7 for a lengthier discussion of career and workplace flexibility.)

Research has consistently drawn attention to the benefits and potential of part-time work for academics (Drago & Williams, 2000). This may be offered as a temporary modification or as an ongoing program to support caregivers and diversity in STEMM. Higher education lags far behind industry in offering part-time options (outside the adjunct market, which has its own challenges) (Wilson, 2008). A study of family-friendly policies in higher education completed in 2007 by the University of Michigan, Ann Arbor, found that only 15 percent of the 189 institutions surveyed had a formal policy allowing professors to work part-time. In many other institutions, part-time work is negotiated on an ad hoc, case-by-case basis (Wilson, 2008). In academia, part-time work has in many instances taken the form of marginalized and contingent labor, but it does not necessarily need to be so. Instead, part-time work could be structured as a valued and viable alternative for those who need the flexibility that receives adequate pay and benefits and allows continued progress toward tenure and other goals. It is important to note that, much like STC policies, there is evidence suggesting potential stigma associated with the use of part-time work that can be harmful particularly to women's career advancement (Durbin & Tomlinson, 2010; Van Osch & Schaveling, 2020). As with stop-the-clock policies, attention needs to be paid to implementation to ensure equitable outcomes for all who need part-time options. (See examples of part-time work from industry and academia in Chapter 7.)

Policies Related to Direct Care Support

Childcare and Adult Dependent Care

Universities and colleges may offer on-site childcare options, but rarely enough to meet the high demand for this resource. Across nine research universities, on-campus childcare was faculty members' most requested family-responsive service in a 2005 study (Ward & Wolf-Wendel, 2005). Still, existing research suggests that this is not a widely available benefit (Forry & Hofferth, 2011).

Access to on-site care is determined institution by institution and varies widely, as does the quality of care offered. Variation in access and eligibility has consequences for students and trainees. Typically, these centers, which also provide care for the children of faculty, have a limited number of childcare slots for the children of student and trainee parents and waitlists are common (Hill & Rose, 2013; Reichlin-Cruse et al., 2021). A national mixed-methods study of postdoc parents' experiences found that only 29 percent of postdocs reported being eligible for on-campus childcare, and even fewer could use it (Lee et al., 2017). Postdoc survey participants noted that campus childcare facilities would not accept young infants, had waitlists longer than their postdoctoral appointment would last, and were unaffordable. Further, postdocs reported that the hours of childcare availability did not match their work schedules, which were longer than those worked by others on campus. More recent data from the National Postdoctoral Association's Institutional Policy Report indicate that 40 percent of institutionally funded postdoc trainees had access to on-site childcare in 2019 and that 16 percent were eligible for subsidized childcare, a decrease from 2016 (Ferguson et al., 2021).

Childcare access has also been a central concern for many graduate students and has been raised as a key needed benefit in labor organizing efforts of graduate students across the nation. Individual university collective bargaining agreements may now cover such support on a growing but still small scale. For example, Harvard graduate students obtained a raise and funds to cover childcare arrangements as part of their union contract (Harvard Graduate Students Union, 2020).

Along with on-site childcare, some colleges and universities offer other policies and programs to support parents in caring for their children, including the following examples. Arizona State University offers faculty consultations with a childcare services coordinator, and the University of Pennsylvania provides a childcare resources web page that directs families to

CURRENT LAWS, POLICIES, AND PRACTICES 77

resources and activities. At Brown University, faculty benefit from backup and emergency care, as well as financial support for the family-related expenses incurred with work travel. The University of Chicago offers travel grants of up to $500 per year for faculty needing to travel for work with their children. The University of Houston has developed summer camps for children, and the University of Pennsylvania provides backup care for snow days (Cardel et al., 2020). Rice University's "Children's Campus" provides childcare for faculty members, with participants reporting that it decreased their job stress (O'Brien et al., 2015).

Beyond childcare, some educational institutions provide services to support faculty and staff in caring for aging or disabled family members. The committee did not identify evidence of this type of care provision for students. Care services may be provided via third-party contractors off site or on site, paid for with university subsidies, or the institution may provide referrals and subscriptions to care-finding services. The committee could find little to no research examining these types of programs in detail and their prevalence at academic institutions, though the relative dearth of information on older adult care is likely suggestive that this is a less common benefit and in line with the greater attention that has been given to childcare compared with other caregiving situations.

Policies Addressing Bias and Discrimination

Colleges and universities have workshops and other initiatives to address bias based on race/ethnicity, gender, and other categories. While many of these measures are not specific to the potential for family responsibilities discrimination, general measures aimed at reducing bias in the hiring process have been employed and may have important implications for caregivers. Key measures universities have implemented to address bias and discrimination include ensuring diverse committees for hiring and promotion, providing training, and using rubrics for evaluation.

There is less information on current policies aimed at addressing the potential for family responsibilities discrimination in academic workplaces, but this type of discrimination is prevalent. Reports across the labor market in fact have risen over time. Between 1998 and 2012, for example, reports of family responsibilities discrimination increased by over 500 percent (Calvert, 2016).

Importantly, family responsibilities discrimination may also be different for different individuals. For example, in 8 percent of more than 4,000 cases, individuals also alleged racial discrimination, such as when

Black employees were denied leave or flexibility given to White caregivers (Calvert, 2016). As such, this is an important area for greater attention among universities, particularly for the potential for compounding forms of discrimination against those with intersecting marginalized identities.

SUMMARY OF FINDINGS FROM CHAPTER 4

A variety of actors shape the specific policies and practices that any caregiver in STEMM will find available to them. Federal and state laws form a patchwork of different policies that provide varying degrees of support and protection for university employees, but they may not provide the needed resources or support for all caregivers. Institutions themselves provide various forms of support for caregivers, ranging from leave to accommodations to childcare to protections against bias. There is, however, no universal standard across universities, and what this looks like in practice varies across institutions. The result of this patchwork of policies is that some caregivers can access the support needed to remain in STEMM fields and thrive, while others have little to no support during times when they are engaged in caregiving.

1. There is no single law that establishes the full rights of family caregivers in the United States. Instead, a disconnected set of mandates and regulations at the federal, state, and local levels provides legal protections and accommodations for family caregivers that can make it challenging for caregivers to understand their full rights and leaves significant gaps in the support caregivers can access.
2. Along with the federal government, other third-party actors influence the support caregivers have access to. Federal agencies and other funders may provide additional funding, no-cost extensions, and other resources to support caregivers. Accrediting agencies may also set caregiving-supportive requirements for universities to meet to receive accreditation or certification.
3. Though legal and accreditation requirements may create some degree of uniformity in policies, there is no universal approach to supporting caregivers at colleges and universities across the United States.
4. Colleges and universities have enacted a variety of policies to support different constituencies, including students, trainees, and employees.

5. The most common sets of family-friendly policies and programs across universities include caregiving leave, stop-the-clock and other extensions, educational accommodations and work modifications, and childcare provisions or subsidies, though how these are implemented vary across institutions.
6. While colleges and universities have done some work to address bias and discrimination based on gender and race/ethnicity in their admissions, hiring, and advancement processes, few if any measures are specifically designed to address family responsibilities discrimination.

5

Barriers to Effective Policy Implementation

This chapter draws substantially from the research paper "A Comprehensive Literature Review of Caregiving Challenges to STEMM Faculty and Institutional Approaches Supporting Caregivers," by Joya Misra, Ph.D., Jennifer Lundquist, Ph.D., and Joanna Riccitelli, which was commissioned for this study.[1]

While many potential policies and programs exist to support family caregivers, a range of barriers to effective implementation remains. Poor policy implementation, unintended consequences, and policies that are implemented but not sustained over time can serve to undermine the efficacy of policies for caregivers in science, technology, engineering, mathematics, and medicine (STEMM). This chapter highlights six barriers to successful policy implementation and use: (1) affordability, (2) availability, (3) lack of awareness, (4) lack of attention to intersectionality, (5) lack of institutionalization, and (6) cultural beliefs and biases. The committee acknowledges that institutional context varies greatly and certain colleges and universities may encounter challenges and constraints not mentioned here. The goal was to focus on issues that cut across contexts and may be experienced by most institutions when implementing caregiving policies.

AFFORDABILITY

While in many countries families receive subsidized or free childcare and other resources, in the United States, families carry the financial burden of care (Garfinkel et al., 2010). Financial costs of care can be a barrier to policy access for individuals in academic STEMM who may not be able to afford the support they need without heavy subsidies. Potential costs can also be a barrier to implementation because universities looking to cut

---

[1] The full paper is available at https://nap.nationalacademies.org/resource/27416.

spending or balance budgets may be worried about upfront costs associated with certain policies.

One example that is frequently central to debates about costs of family caregiving in the United States is the provision of paid care, such as childcare or adult care centers or home aids. While universities may offer subsidies or provide on-site childcare (adult care is much less common) and some funders offer grants to cover some costs associated with care, financial barriers can impede how effective these resources are for individuals, as well as the likelihood of universities instituting or expanding support. First, looking at individual affordability, existing care on college campuses is rarely subsidized, and the cost often exceeds what the U.S. Department of Health and Human Services designates as affordable given the average faculty salary (Dolamore et al., 2021). Groups within STEMM who are outside the tenure track, such as students, trainees, and staff, can struggle even more without subsidized support given their income levels. In a letter sent to the National Institutes of Health detailing challenges faced by postdocs with children, the letter writers analyzed data on average housing and childcare costs in cities across the country. Their results found that the cost of rent and care often exceeded 30 percent of income in a two-postdoc household, and in extreme cases, amounted to more than 75 percent of income, especially for those with more than one child (Guo et al., 2023). The challenges of affordable care also extend well beyond universities, as costs for childcare and long-term care are soaring nationwide (Abelson & Rau, 2023; DeParle, 2021). Given all of this, without affordable care, the provision of care alone is not enough because many in academic STEMM will not be able to access what they need within their means.

At the same time, providing and subsidizing childcare represents a large upfront cost for universities, and this may hinder willingness and ability to provide as much support as caregivers need. This is especially challenging in a setting where there have been declining public investments in universities, creating a greater pinch for those institutions that are already financially constrained (Marcus, 2019). Unfortunately, limited peer-reviewed data exist on how university administrators make decisions about policies and programs to support caregivers based on budgets and costs. Given this limited literature, the committee sought input from those at the forefront of implementing these kinds of programs through conversations with Sherry Cleary, former dean of the Office of Early Childhood Initiatives at the City University of New York, and with leadership from the National Coalition for Campus Children's Centers.[2] All emphasized the unique value of on-site campus childcare as a powerful recruitment, retention, and research tool. They noted that when programs were reliant on single sources of funding, they were especially at risk. Dr. Cleary specifically noted: "Programs have to demonstrate their integral alignment to the mission of the university, and if they can't provide detailed cost-benefit analyses of this, essential service programs can be at risk when campus budgets are challenged."

Ultimately, while policies like on-site care and subsidies can have long-term cost-saving benefits for universities and are important for individuals to be able to continue their work and education in academic STEMM, the costs of supportive care policies can pose a barrier to access, implementation, and expansion.

AVAILABILITY

Availability is another key barrier to policy efficacy, as existing policies may be limited in scope in ways that present challenges for family caregivers. The provision of paid care is again a useful example of these issues. On-site childcare would help students and academic STEMM faculty and staff access quality care for their young children at or near their workplace, yet on-site childcare is rare among private U.S. employers (Galinsky et al., 2008), and no representative survey of colleges and universities has assessed how many campuses provide childcare subsidies or on-site care. However, recent surveys of specific academic fields suggest the number may be higher in some areas of academia, but coverage is uneven (Dolamore et al., 2021; French et al., 2022). Additionally, an analysis of community colleges and public universities in 36 states found substantial declines in campus childcare between 2002 and 2015 (Eckerson et al., 2016). Moreover, when on-site care is available, full-time slots are not always offered (Dolamore et al., 2021). And many care centers have extensive waiting lists given much higher demand than they can meet. Significantly, research suggests that the availability of on-campus childcare has been declining at the same time that the population of students with children is increasing across all institution types, compounding challenges of availability (Noll et al., 2017).

---

[2] These conversations took place over Zoom on July 3, 2023, with Dr. Cleary and on July 13, 2023, with representatives from the National Coalition for Campus Children's Centers. The committee engaged these experts given their extensive experience in building and growing childcare centers on university campuses and the challenges of doing this.

Issues of childcare availability came up frequently in interviews as well. Indeed, the inadequacy, inaccessibility, or effective unavailability of institutional childcare supports was a subject of widespread critique among the study sample—perhaps because interviewees so desperately needed these services, as articulated in the quotes below:

> "I know it's a really good day care center, but it has a waiting list 6 years in advance. So, you essentially have to know before you get pregnant that you're going to use this service, and it costs a lot of money."

> "Childcare [would be helpful], yet that's privileged access. I could only get that for my second child, once I was faculty. It's super tough if you're not faculty to get the privilege of affordable childcare. There is a barrier, and the institution would say, "We just can't provide it for everybody, and there's always a 2-year waiting list." [But] having access to affordable care would go a long way for a lot of junior faculty and even graduate students or postdoctoral fellows."

Of course, on-site childcare is just one of many availability issues regarding care resources. Childcare provisions may still be less robust than necessary, but older adult care is even less common. The committee could not identify any comprehensive reviews of the provision of adult care on campuses. In addition, finding examples of this in practice proved difficult, though they found one promising practice for providing adult care at Virginia Commonwealth University (see Chapter 6). Given the relative lack of attention to the needs of caregivers of adults relative to caregivers of children, a key gap remains in availability of the kinds of supports caregivers of adults need that those caring for children may be better able to access. In interviews, many respondents spoke of challenges with access to older adult care. As one respondent noted:

> "Many institutions have a childcare benefit that allows you to place your child in daycare, or they have a daycare themselves, or they have some childcare benefit. I'm finding it difficult to find eldercare benefits. So, it becomes really challenging."

LACK OF AWARENESS

Lack of awareness of existing caregiving institutional policies remains a key barrier to use and efficacy of family-friendly policies in academic

STEMM. No matter how beneficial or supportive a policy or practice might be, it cannot have its fully intended effect if people are not aware of its existence, their rights to access it, or how to access what they need (Calvano, 2013; Dembe & Partridge, 2011). Multiple studies have identified a lack of awareness about caregiver policies among STEMM employees. For example, in a 2015 study of faculty at a large research institution, 91 percent indicated they were unaware of their university's policies and procedures for older adult care (Leibnitz & Morrison, 2015). Similarly, a 2023 study of ophthalmologists found that three-quarters of those surveyed did not know whether their workplaces had stop-the-clock policies for tenure (Kalra et al., 2023).

One reason for lack of awareness may be simple challenges of knowing where to find information on particular policies and eligibility criteria. In the 2015 study of faculty at a large research institution noted above, many respondents reported difficulty finding the information they needed (Leibnitz & Morrison, 2015). (Chapter 6 provides examples of institutions that have developed websites and offices that make it easy to find existing resources.) Another barrier to awareness and access comes from supervisors and department chairs, who often serve as gatekeepers to policy access even when they are not required to approve a particular benefit (Shauman et al., 2018). Chairs and supervisors taking on this role are themselves not always aware of what policies stipulate and may misdirect those they are intending to guide. Shauman et al. (2018) note that this underscores the need to train department chairs and other supervisors so that they are aware of policy details and why facilitating access to these policies is so important.

Lack of awareness could have substantial consequences, as evidenced in interviews. Students and other early-career scholars reported great difficulty accessing even basic caregiving supports, such as postdelivery recovery time and parental leave or tenure clock adjustments because of awareness issues. It was not unusual among interviewees to learn of available formal supports after the point at which it would have been helpful to access them, leaving them completely without support. Students often did not know what they were entitled to or were reluctant to ask. As one respondent shared:

> "I've spoken with other people who had children earlier on [as students] and I think most of them, in my experience, had to navigate it themselves.... I just think there needs to be some kind of standard for that [returning to class after giving birth] from like a normal recovery standpoint."

And interviewees also recounted challenges created because of a lack of awareness among those who were supposed to be guiding them. In some cases, administrators were not sufficiently knowledgeable about what was available or permissible to be helpful even when approached for assistance. For example, a graduate student who left academic STEMM due to the financial pressures she faced as a caregiver for her grandmother recounted:

"With the additional stipend, I think it was like [whether] you have a dependent. I think the real stymie with that one was … not understanding whether or not they had to be my dependent on federal taxes or something like that. And the person who was in charge of dealing with that also didn't know the answer to that and they just weren't really all that informed on their own policies because it didn't come up that often."

Any policy can be effective only if it is used, and lack of awareness can pose a substantial hindrance. Without clear documentation online and clear communication to department chairs and others who may serve as gatekeepers to policy access, students, faculty, and staff can remain uninformed about the support available to them. And, as is discussed later in this chapter, cultural beliefs and biases can lead to a fear of being stigmatized for utilizing caregiving resources, resulting in concerns among caregivers in reaching out to receive the information they need (Shauman et al., 2018).

LACK OF ATTENTION TO INTERSECTIONALITY

Policy effectiveness is also hindered by a lack of attention to intersectionality, which can lead to unintended consequences that leave out women of color and other groups with intersecting marginalized identities. Because Black, Hispanic, and Native women as well as LGBTQ+ women or nonbinary people often represent incredibly small numbers across academic STEMM disciplines, their experiences and needs frequently get lumped together into the broader category of women or dropped from analysis. Doing this, however, ignores the heterogeneity among women and has the potential to lead to policies that may be effective for White women, but are ineffective or insufficient for others (McAlear et al., 2018). As Kossek, Lautsch, et al. (2023) argue, intersectionality means that even if institutions put in place some caregiving supports, some groups may not be able to access these supports, with these effects particularly salient for those from more disadvantaged groups.

For example, definitions of "family" influence how leave policies are written and implemented. As noted in Chapter 3, family caregiving is not the same for everyone, and it varies by race/ethnicity as well as immigrant and LGBTQ+ status. While White Americans are more likely to provide care for those within the nuclear family unit, immigrants and Americans of color are much more likely to provide care for an extended network of family, kin, and community (Gerstel, 2011; McCann et al., 2000; Sodders et al., 2020; Tam et al., 2017). Research has shown this broader definition of family is also more common among LGBTQ+ individuals (Biblarz & Savci, 2010; Weston, 1997). When policies and programs are constructed around White, heteronormative assumptions of a narrow definition of family caregiving, this presents challenges for those who do not subscribe to this. For example, caregiving leave policies are often provided on an individual, ad hoc basis (Roselin et al., 2022), which leaves them open to potential bias and normative assumptions of who needs such policies and for what purpose. Policies may be written such that leave is granted only to those caring for the needs of an immediate family member. And even if policies apply to a broader definition of caregiving and are being shared by well-intentioned department chairs advising faculty members or faculty advisors assisting students, chairs and advisors may still implicitly assume that family leave is not as expansive. This can hinder access for those taking on care responsibilities of individuals to whom they are not directly related. Attention to intersectionality is necessary to ensure policies, programs, and resources have the greatest effect for all those who make use of them.

In interviews, caregivers of color frequently discussed the ways in which implicit assumptions and biases from the dominant White culture clashed with how they understood family caregiving in ways that created greater challenges for them and produced barriers to feeling fully supported by present policies. As one caregiver of color noted:

> "It's very typical and in keeping with my culture to take care of your elders. And so, I always knew that this responsibility was going to fall on me. I had learned of [my mother's] diagnosis, and I knew then that I was no longer going to be able to pursue the career that I wanted. [After my mother died], taking on my grandmother was not a question of, like, do I want to? It was just, like, it's a natural thing.… Obviously, I'm aware that, like, White

culture isn't like that [but for me] it's, like, what was expected and what's normal."

Along with failure to fully support caregivers with intersecting marginalized identities, a lack of attention to intersectionality can fail to account for all possible care recipients. An example was shared in interviews where a parent discussed the challenges of trying to utilize policies that were not made with children with disabilities in mind:

"The university is really proud of the fact that they have this relationship with a company ... that's supposed to help you find care, and so if something comes up and you've got a kid that's sick, they're supposed to be able to find you a last-minute babysitter so you can still get to your work activities and your kid can be taken care of. They love this idea.... It's a bizarre policy in some ways. And [when] I have called them ... they can never find me a caregiver that can take care of a special needs kid. So those of us with kids with autism or developmental issues or behavioral issues or severe ADHD [attention deficit hyperactivity disorder], we can't use that program anyway.... People think they're helping, but they're not really helping."

While the policy attempted to provide paid caregiving support to help parents manage a child's sickness, implicit assumptions that did not account for those sick children also having disabilities or developmental challenges left the parents of these children without access to support.

LACK OF INSTITUTIONALIZATION

Another barrier to ensuring access to effective policies for family caregivers is ensuring that effective policies remain in place. There has been less academic research in this area, as much of the evaluation literature focuses on how policies operate when implemented, not what happens after implementation to ensure policies remain. Here, the committee drew on examples to inform its discussions and highlight how this can hinder positive interventions.

One such example is the time-banking system instituted by the Stanford University School of Medicine (discussed in more detail in Chapter 7). The program, which was started as a pilot in 2012 across several clinical and basic science departments, allowed medical faculty to "bank" time spent on work such as mentoring or covering another colleague's shift that is not typically as highly valued in the field. This banked time could then be used as credit to get back time for academic activities including grant writing support, but most importantly for family caregivers, for "home-support activities" (Berg, 2018; MacCormick, 2015). Along with piloting the program, researchers also conducted a multiyear study to assess its efficacy. The study found many benefits to the program, namely, improved perceptions of flexibility and wellness as well as greater institutional satisfaction. The authors also found that compared with a matched sample of nonparticipants, those who participated in the program received 1.3 times more grant awards, totaling around $1.1 million in funding per person (Fassiotto et al., 2018). Despite these benefits, however, the program has been retained only in the Department of Emergency Medicine as of 2022. A policy established under one dean that was successful across each of these different metrics, in both employee perceptions and grant outcomes, was not successfully institutionalized and, as a result, was discontinued after that dean stepped down.

To the committee's knowledge, no research currently exists explaining the reasons why this policy or others like it did not continue; however, some literature examines sustainability of organizational change that can provide insight into why and how it is that policies either remain in place or fade away. The forces affecting policy staying power operate at multiple levels and encompass factors ranging from issues of timing, leadership, managerial support, individual challenges, cultural conflict, organizational and procedural barriers, and outside influences from external events (Buchanan et al., 2005).

One of the key barriers to policy staying power, however, is when a policy fails to become integrated into the structure of an organization, that is, when it fails to become institutionalized. Policies tied to groups or individuals who champion them rather than to institutional units or positions that exist beyond any individual are less likely to remain intact, especially if their champions leave the organization (Wynn, 2019). Even policies that have demonstrated positive outcomes, positive media attention, and interest from other organizations looking to implement something similar have been documented to phase out following the departure of their main backer when they were not effectively built into the structure of the institution (Wynn, 2019).

There is no guarantee that effective policies remain, and anecdotal evidence in universities and research in other domains suggest that this can be a substantial challenge. In particular, without institutional support and ensuring that policies and programs become a part of the structure of the university, good policies may not last as long as they should.

CULTURAL BELIEFS AND BIASES

Some barriers are grounded in deeply embedded systems in colleges and universities. As noted in Chapter 3, academic STEMM has a core cultural assumption that single-minded devotion to work is an indicator of scientific merit, and this cultural assumption is institutionalized into many standard policies and practices, including full-time, time-intensive tenure clocks and productivity metrics that do not consider periods spent focused on caregiving (Blair-Loy & Cech, 2022; Blair-Loy et al., 2023). A challenge to the introduction of new policies is ensuring cultural change to go along with it, as this helps them to become embedded within the institution. It also helps ensure people do not fear stigma or backlash or face biased and discriminatory treatment for using them (Kachchaf et al., 2015; Kossek & Lee, 2022; Sallee, 2012; Williams, 2005, 2014; Williams & Norton, 2010).

Although flexible work scheduling norms have not made the same inroads into the academic setting as they have in some firms and corporations outside academia, such arrangements have made a major difference for the success of employees in these other settings (Christensen, 2013). As such, scholars have suggested variations within academia, such as the half-time tenure clock that operates on a 12-year track (Drago & Williams, 2000; Moors et al., 2022; Ward & Wolf-Wendel, 2012). Certainly, the precedent of the pandemic has sparked a shift in the way that tenure delays are viewed and accepted as well as instigated tenure clock extensions for all faculty at many institutions (Smith et al., 2022). However, so long as standards for and measures of success in academia remain unchanged—for example, a reliance without question or contextualization on metrics known to have bias, such as citation indices, as well as excessively high expectations of publication numbers—and ideal worker norms and expectations of workplace devotion remain entrenched, caregivers who opt for these pathways will continue to fall behind in academia. And, indeed, many faculty who have engaged in these practices have been disadvantaged in terms of promotion and tenure (Williams & Norton, 2010). Ultimately, cultural biases and flexibility stigma create a disincentive to use flexible policies, resulting in the uneven uptake seen among STEMM faculty (Lundquist et al., 2012; Morgan et al., 2021; Sallee, 2012). Concerns about this stigma may be part of why most faculty members on modified duties or parental leave continue to engage in research and mentoring if they can (Lundquist et al., 2012).

Cultural change is challenging; however, the COVID-19 crisis put into motion a series of shifts that challenged conventional practices in higher education. Indeed, some used it as an opportunity to recalibrate the approach to the standards used to evaluate productivity and effect. As one example of broader efforts to promote cultural shifts during the pandemic, a team of researchers called on the field to reinvent promotion and tenure practices by, for example, advocating for the use of alternative impact metrics, such as "communication, community-based implementation, dissemination (e.g., Altmetric scores), effective mentoring, and advocacy work" (Cardel et al., 2020). Yet, while the pandemic shifted the paradigm in several ways, the innovations universities adopted have so far failed to fundamentally change the criteria for tenure and promotion. In a study of pandemic policies, less than 1 percent of the top 386 U.S. universities modified their tenure and promotion evaluation expectations in some way (e.g., to note that quality over quantity would be considered for evaluation or that other duties besides research would be given more weight) (Culpepper & Kilmer, 2022).

Cultural beliefs and biases can discourage use of available work-life supports, including more quotidian ones like paid time off, as evidenced in interviews:

> "I'd say fear of stigma and discrimination … some people may be fearful of utilizing the work-life support policies that are put in place because such expressing a need for work-life balance could lead to negative career consequences.… Also, there is the workload and time constraints where very heavy workloads and also time constraints can make it challenging for us to take advantage of work-life support policies. Maybe you have very high demands, you have tight schedules and tight deadlines, so it's very rare for you to even ask for a day off."

This highlights the steep barriers cultural biases create for policy success. Even in the face of a global pandemic, entrenched norms are still held in many instances. It may take time and creativity, but effective policy implementation needs to work to break down these norms and develop new norms that value flexibility and recognize and embrace the outside lives of those in academic STEMM, including their caregiving responsibilities. Efforts that normalize flexibility and outside obligations through support for institutional and departmental leaders, greater visibility and acknowledgment of caregiving, and policies that center work-life inclusion are important for overcoming cultural barriers.

SUMMARY OF FINDINGS FROM CHAPTER 5

Policies that provide needed support to caregivers are not always as effective as they could be and may even fail to be implemented or sustained. Understanding these barriers is crucial to designing effective policies that consider potential pitfalls and aim to address them. In Chapter 6, we draw on knowledge of these barriers as well as existing evaluative research to outline foundational and promising practices to support family caregivers in academic STEMM.

1. Even effective policies may fail to be implemented, expanded, or continued given barriers that affect utilization, awareness, and practicality.
2. The upfront costs associated with certain policies, such as childcare provisions, can be a barrier not only to individuals who may not be able to afford quality outside care but also to institutions looking to add or expand offerings for caregiving on campus.
3. Availability of paid care is a challenge for many in academic STEMM and many Americans more generally. There have been declines in care centers on campuses over time while there has been a rise in certain populations, such as students, with children, creating even greater availability challenges for on-site childcare options.
4. Across multiple kinds of policies, people are simply not aware of what their universities offer, making it harder for them to make use of policies and programs they could access.
5. Without considering intersectionality, policies may produce unintended consequences that leave out women of color, socioeconomically disadvantaged, and other groups whose experiences and definitions of family caregiving do not fit dominant norms and assumptions.
6. Cultural biases, particularly flexibility stigma, can make it difficult for people in academic STEMM to utilize family-supportive

policies for fear they will be seen as less committed and dedicated to their work.
7. Policies that are not fully institutionalized but instead championed by one person or one group risk being discontinued if the policy champions leave the organization.

6

Best Practices for Colleges and Universities

This chapter draws substantially from the research paper "Comprehensive Literature Review of Current and Promising Practices to Support Unpaid Caregivers in Science, Technology, Engineering, Mathematics, and Medical STEMM," by Jessica Lee, J.D., Erin Frawley, M.Ed., and Sarah Stoller, Ph.D., which was commissioned for this study.[1]

The report so far has outlined the existing family caregiving landscape, and the many challenges caregivers face as well as barriers to successful policy implementation. In the next set of chapters, the committee turns toward action. It is not simply enough to understand the reality of family caregiving in academic science, technology, engineering, mathematics, and medicine (STEMM); there is an immense need for concerted effort to address these challenges and provide greater support to not only help individual caregivers but advance equity and inclusion in STEMM and ensure the continuation of a strong and supported workforce to advance STEMM innovation. While there is no one-size-fits-all model, there are data and evidence to guide the kinds of approaches colleges and universities can take to implement effective policies that provide caregivers with the support they need. This chapter begins with a discussion of the foundational minimums that colleges and universities must meet to ensure legal compliance, given many reports of institutions that are noncompliant with existing laws, and considers policies addressing bias and discrimination. From there, the committee details current knowledge of the best practices for each of the remaining three policy areas covered by current institutional approaches as outlined in Chapter 4: leave, accommodations and adjustments, and direct care support. The committee defines best practices as those with a body of literature examining their effectiveness as well as evidence of application

---

[1] The full paper is available at https://nap.nationalacademies.org/resource/27416.

and feasibility in a college and university setting. This is distinct from the innovative practices detailed later in Chapter 7, where there is more limited empirical evidence to support policy effectiveness and/or the policies have only been implemented in other domains. The chapter concludes with a discussion of practices to challenge existing cultural norms given the significant barrier this can pose to effective policies. Throughout the chapter the committee offers checklists for key actions to implement best practices based on committee expertise and review of the literature (see Boxes 6-1, 6-3, 6-5, 6-7, and 6-10 for best practices check-list) as well as call-outs of examples in action showcasing practices at institutions around the country (see Boxes 6-2, 6-4, 6-6, 6-8, 6-9, 6-11, and 6-12 for examples in action).

As evidenced in Chapter 5, even best practices can fail or face unintended consequences due to various structural and systemic barriers. Though these practices remain important and needed to support family caregivers, attention should always be paid to thoughtful implementation and evaluation to assess the true effect of new policies and procedures. Individual institutions should examine how any new practice works in their context and adjust as needed.

FOUNDATIONAL MINIMUMS FOR LEGAL COMPLIANCE

The minimum best practice for federal, state, or municipal laws affecting caregivers is compliance with the laws. But this is not always easy, as the legal regime surrounding caregiving is complicated. As there are many reports of instances where universities are not in full compliance with existing laws and regulations protecting family caregivers, a necessary starting point for any discussion of best practices for supporting family caregivers is ensuring a grounding in what institutions are legally required to do (Calvert, 2016; Gulati et al., 2022; Lee et al., 2017; Mensah et al., 2022; Williams et al., 2022). For more on these existing laws and regulations, see Tables 4-1 and 4-2 in Chapter 4, which provide an overview of legal mandates for both employees and students.

Effective Compliance with Laws Governing Students

Starting with Title IX of the Civil Rights Act of 1964, there are several areas where compliance and implementation of the law on college campuses may be lacking. Principally, Title IX prohibits discrimination based on sex in all educational programming. The ban includes discrimination on the basis of pregnancy, but this has often been ignored by many institutions until recent years (Mason & Younger, 2014). Colleges and universities need to ensure their policies do not discriminate against pregnant students, notably by failing to provide as much leave as their medical provider says is medically necessary and by failing to provide pregnancy accommodations such as the ability to avoid exposure to toxics during laboratory research. More broadly, they need to create an environment where faculty and other university staff understand caregivers' legal rights. Along with ensuring basic compliance, following these requirements to support student parents is also an issue of racial equity, as the majority of undergraduate student parents are students of color (Institute for Women's Policy Research & Aspen Institute, 2019).

To maximize the effectiveness of federal laws that prohibit caregiving-related discrimination and bias, operationalizing key aspects (and enforcing them) at the institutional level is extremely important. Reuter (2006) calls for the enforcement of a "strict" policy to that end. Taking implementation of Title IX protections against discrimination as one example, and as noted in the legal compliance checklist in Box 6-1, a best practice is to ensure that anywhere Title IX is mentioned on campus communications materials, it should be highlighted that pregnant and parenting students have rights. Web pages and other communication materials also should be inclusive and representative of a wide variety of caregivers and parents (The Pregnant Scholar, 2022).

Though the 2023 Title IX regulations have yet to be released, it is also important for universities to understand future issues of compliance. One area is the provision of lactation accommodations for students. The new Title IX regulations once enacted include a clear requirement for educational institutions to provide their students with a clean, private, non-bathroom lactation space and the time to use it. To avoid negative consequences on a student's education caused by missing class, universities should strive to place lactation rooms in areas readily accessible to the students who need them. Because students often struggle to arrange class schedules around their lactation breaks, supportive institutions often have a lactation policy making clear to faculty that these students should be excused without penalty (Clark et al., 2021; The Pregnant Scholar, 2020).

Additionally, the U.S. Department of Education's draft revised Title IX regulations will require training of all employees on Title IX[2] (Office for

---

[2] These revised Title IX regulations were expected to be enacted in October 2023 but have been delayed.

BOX 6-1
Checklist for Training to Create an Environment Where University Staff and Leaders Understand Caregivers' Legal Rights

- ☐ Ensure that Title IX officers understand that Title IX requires birth-parent leave and pregnancy accommodations for students.
- ☐ Designate a point of contact in human resources, and in the provost's office, so faculty and staff with issues related to leaves and accommodations receive accurate information regarding their rights. Provide the point of contact with training to ensure that matters are handled by someone who understands the complex and overlapping legal requirements.
- ☐ Train department chairs and individual faculty about the rights of pregnant and parenting faculty and students, including birthing students' rights to leave as long as medically necessary.
- ☐ Include discussion of "maternal wall" bias against mothers, and against fathers who engage in family caregiving in all antibias trainings, orientation trainings, department chair trainings, and the like.
- ☐ Train faculty and administrators so that they know that it is illegal to make someone "pay back" a leave before (or after) they take it.
- ☐ Train department chairs and faculty so that they know it is illegal to require anyone on leave (paid or unpaid) to work, to discourage anyone from taking leave, or to penalize anyone for doing so.
- ☐ Ensure that Title IX officers, faculty, and other officials know that students, faculty, and staff are entitled to pregnancy accommodations, and have a thorough command of the types of accommodations that are workable.
- ☐ Ensure that Title IX officers, faculty, and other officials know that students, faculty, and staff are entitled to a clean and accessible place to pump milk that is not a bathroom, and that students should not be penalized for being late to class if the only lactation space available is so remote that the student cannot arrive to class on time.
- ☐ Ensure that faculty know that it is illegal to treat caregivers less flexibly than others who need time off and accommodations due to nonwork responsibilities and commitments.
- ☐ Include in all sexual harassment training the legal requirement not to harass students, faculty, and staff based on pregnancy and related gender issues.
- ☐ Share Title IX protections widely and with an inclusive representation of a diverse array of caregivers.
- ☐ Make caregiver-friendly policies that apply to all caregiving responsibilities and are automatic rather than narrowly applied to women/mothers and opt-in requirements.

Civil Rights, 2022). Such training is especially important as the efficacy of Title IX offices is often hindered by a lack of knowledge and awareness of these protections. A recent study found that only 35 percent of those surveyed were aware of the Title IX office and resources offered. The new Title IX regulations require educational institutions to disseminate necessary information via their Title IX coordinator's programming and grievance policies (Office for Civil Rights, 2022).

Effective Compliance with Laws Governing Faculty and Other Employees

Along with ensuring Title IX compliance for students, universities may need to address important compliance issues for faculty and staff. For example, it has been documented that various residency programs require individuals who take leave to work extra to "pay back" their leave before they take it (Gulati et al., 2022; Peters & Hartigan, 2023). In addition to possibly increasing the rate of pregnancy complications, this practice violates the Family Medical and Leave Act's (FMLA's) prohibition on denying legally protected leave or penalizing individuals for taking leave. It may also be an illegal form of sex discrimination when other employees are not required to make up work hours expected to be missed for incapacitation and serious health needs not related to pregnancy. It is essential that institutions stop forcing residents, physicians, and other employees to work additional hours to "make up" for their anticipated maternity leave (Gulati et al., 2022; Peters & Hartigan, 2023). Institutions need to realize that since those practices violate the FMLA, they are illegal.

Several mechanisms have been identified to provide funding for leave-talking in medical schools. Most promising is the University of California, San Francisco, policy, which provides for paid leave for faculty, financing it through a small contribution (less than 1 percent of salary) made by faculty for this purpose (University of California, San Francisco, 2020). As an alternative, Gulati et al. (2022) suggest that hiring locums, or a professional who can fill the role of a colleague for a temporary period, is one approach to manage anticipated staff demands associated with leave. Institutions may also want to consider providing support in negotiating a plan and coverage for duties during leave (Cardel et al., 2020).

Relatedly, it is illegal to require or pressure faculty to perform work during leave or to penalize someone for taking leave they are entitled to. A common finding is that faculty employees continue to work while on leave (Ollilainen, 2019; Schimpf et al., 2013). Requiring an employee to work while on FMLA leave may constitute illegal interference with that leave, opening institutions to liability (Gulati et al., 2022). Studies have found that employees' leave is viewed more favorably when they remain engaged while away, which can skirt close to or over the line of requiring work while on leave (Ollilainen, 2019). To avoid not just legal risk but also employee burnout, institutions should consider leave planning that sets clear terms and limits on any work being done while away and ensures that employee time working while on leave is tracked and not deducted from their leave allowance. For example, an employee who desires to continue to check work email and participate in decision-making may be able to take 14 weeks considering time spent working, rather than 12 weeks if no time was spent working during leave.

As of December 2022, new legal requirements are also in effect to provide as-needed lactation breaks and a private, non-bathroom lactation space for all workers who need it (U.S. Department of Labor, 2022). This law may ease some of the challenges long faced by lactating employees in STEMM in the absence of legal protections (Sattari et al., 2020; Shauman et al., 2018; Soffer, 2019); federal workplace lactation law previously excluded salaried and professional workers, such as faculty and physicians, leading many to treat providing lactation space as a special "favor" to women. This is especially important in a climate where many physician mothers report experiencing breastfeeding discrimination (Jain et al., 2022; Ortiz Worthington et al., 2019; Shauman et al., 2018).

Caregivers and pregnant employees also have legal protections against discrimination. One mechanism through which universities can aim to reduce discrimination is through making policies universal rather than contingent on a person's position or identity and automatic rather than opt in. Recent research examines the case of tenure clock extensions. Based on an analysis of 508 universities, the authors find that universal, opt-out tenure clock extension policies predict an increase in the representation of all groups of women among tenured faculty following adoption. In contrast, tenure clock extensions available only to women or that are opt in increase the share of tenured White women but do nothing for women of color (Gonsalves et al., 2022). This underscores one way to address the intersectional nature of family responsibilities discrimination discussed in Chapter 4 and the barriers posed by lack of attention to intersectionality in Chapter 5. By making policies universal and automatic, there is less opportunity for faculty of color to be denied leave and other flexibility granted to White caregivers (Calvert, 2016).

Universities can also employ bias trainings and workshops and engage in self-studies to assess whether their programs are having the desired outcomes. At a minimum, these workshops should cover bias against mothers, which is the strongest form of gender bias, and should inform participants that it is illegal to penalize fathers for taking parental leave (Correll et al., 2007). This material should be included not only in the basic antibias training but also in specialized trainings, such as trainings for department chairs and search committees. In addition, orientation trainings for new faculty should provide the basic information they need to know about treatment of pregnant and parenting students.

Finally, institutions must also consider that while many pregnant, postpartum, and caregiving employees may need changes in their work duties to protect their well-being, others do not. Supervisors may remove their pregnant employees from high-prestige positions in a misguided effort to keep these employees safe. This has been a particular concern in STEMM fields (Anderson & Goldman, 2020; Englander & Ghatan, 2021; Gulati et al., 2022). Forcing an employee to take an accommodation they do not want or to take leave when they are able to work with adequate protective equipment or other accommodations can be a violation of federal law (Equal Employment Opportunity Commission, 1964; U.S. Congress, 2021). Best practice is to provide a clear point of contact for the employee and to engage with them in an interactive process to identify accommodations that do not adversely affect their career.

BOX 6-2
Examples in Action: Make Policies and Resources Easy to Find and Access

One challenge to effective policies is the ability to find and access the resources that caregivers need and have available to them. The University of California (UC), San Diego, provides links to all family resources on one convenient web page that is easily accessible on its main website. The resources are also divided into categories based on the needs they address, whether for expectant parents, childcare, older adult care, and parental mental health. Resources can be found at the following link: https://blink.ucsd.edu/HR/services/support/family/index.html. UC San Diego also provides links to local and state resources (https://blink.ucsd.edu/HR/services/support/family/eldercare/local.html) as well as national resources (https://blink.ucsd.edu/HR/services/support/family/eldercare/national.html).

Legal Compliance and Concern About Stigma and Retaliation

Along with the various practices needed to ensure greater awareness of existing laws and appropriate implementation, colleges and universities also need to consider potential barriers to reporting discrimination—particularly fear of retaliation. For policies to be most effective, colleges and universities need to consider these potential barriers. Other work has suggested that some ways to reduce fear of retaliation include providing clear and explicit language on actions taken to prevent bias and discrimination, establishing anonymous online reporting systems to safely report misconduct, and using independent committees to investigate complaints (Torres et al., 2023a, 2023b).

BEST PRACTICES FOR INSTITUTIONAL POLICIES

Policies Related to Caregiving Leave

Adopting a formal policy for caregiving leave with clear standards is a key mechanism by which institutions will make their policies easier for students, faculty, and staff to access, and for faculty and administrators to manage consistently and fairly (Bye et al., 2017; Daskalska et al., 2022; Kraus et al., 2021; Roselin et al., 2022). Yet, many institutions continue to provide leave on an individualized, ad hoc basis, a practice that can result in varied outcomes, including bias or other illegal practices (Roselin et al., 2022). Reducing heterogeneity across departments and individuals can ensure a more consistent application of leave and the ability to directly stipulate an expansive definition of family caregiving rather than implicitly assuming that leave for caregiving is or should only be for the care of nuclear family members. Clearly stipulating a broad definition of caregiving to include a wider community network is important to acknowledge and support the broader conceptions of family among communities of color and LGBTQ+ communities (Biblarz & Savci, 2010; Gerstel, 2011; McCann et al., 2000; Sodders et al., 2020; Tam et al., 2017; Weston, 1997). This can also help to address potential bias as detailed in one study of queer faculty mothers, where the author found queer mothers were less likely to be offered formal, paid leaves, potentially as a result of narrow assumptions that only a birthing parent would take leave (Stygles, 2016).

As institutions aim to formalize the leave policies and processes, other factors are also important to keep in mind. For example, gender-inclusive

> **BOX 6-3**
> **Best Practices Related to Caregiving Leave Checklist**
>
> ☐ All policies related to leave should be written out. For students, policies should indicate whether taking leave will require students' training period to be extended.
> ☐ Title IX requires students to be provided birth-parent leave as long as medically necessary.
> ☐ Provide students (including medical students) with 12 weeks of paid caregiving leave.
> ☐ Legal mandates require paid birth-parent leave for all employees to (at a minimum) the same extent disability leave is provided through the institution's disability policy for reasons other than pregnancy.
> ☐ Provide paid birth-parent leave for both faculty and staff rather than relying on disability coverage or on ad hoc arrangements. Start with state-provided paid leave (if available), and add as much additional paid leave as feasible.
> ☐ If paid leave is only available to employees who have opted into an institution's disability system, ensure that staff (and faculty, if applicable) understand before they decline disability coverage that this will preclude them from paid birth-parent leave.
> ☐ The Family and Medical Leave Act requires that eligible employees be provided with at least 12 weeks job-protected caregiving leave.
> ☐ Design leave policies that are available to anyone with a demonstrated need (not just for mothers, or parents) who certifies that the time will be spent on caregiving, to clarify that this is not a free research leave. Message clearly that using family leave as a paid research leave is inappropriate.
> ☐ Provide clear and comprehensive information of what leaves are available, and how to apply for them, on a well-publicized website.

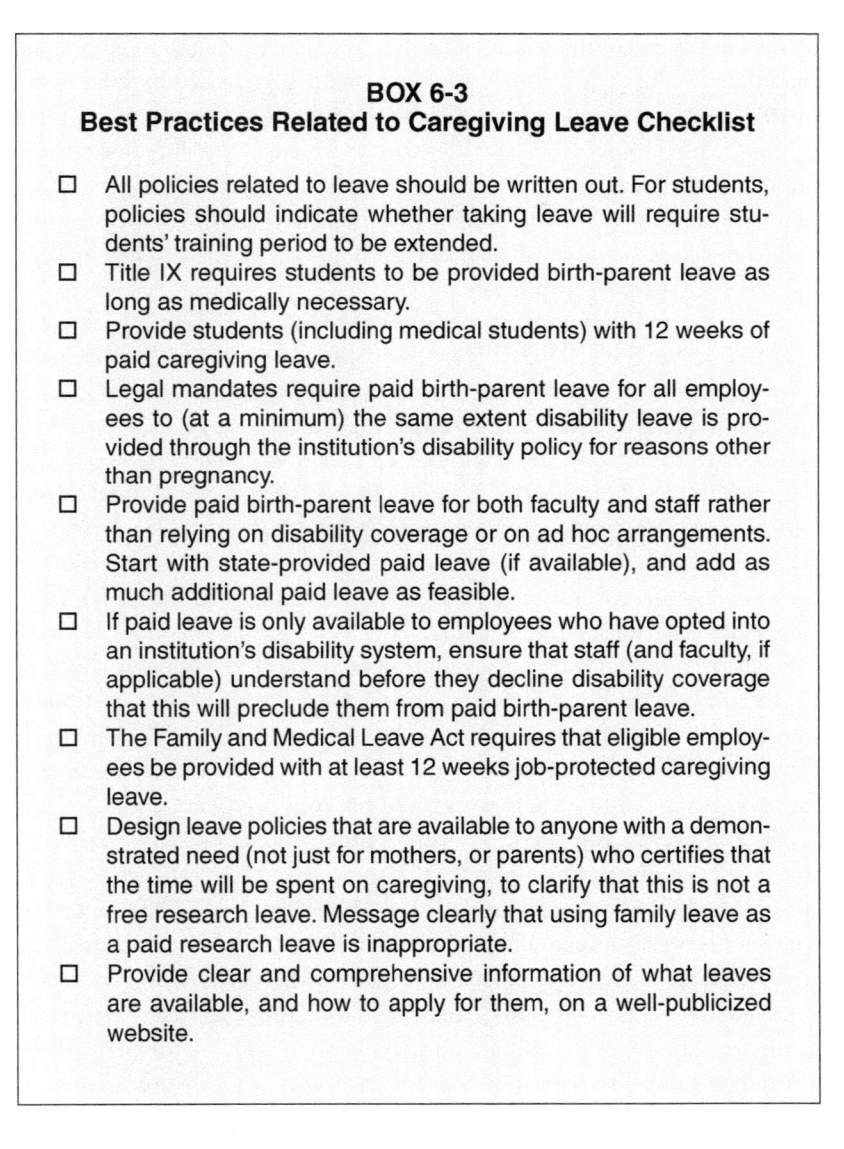

leave policies allow for parents or caregivers of any gender to access the leave they need, but they may cause unintended consequences. Researchers have expressed concern that gender-blind leave benefits may hurt women, especially women faculty, who are less likely to use leave periods to further their academic work than are men (Burch et al., 2023; Feeny et al., 2014). The primary users of these policies remain women who give birth, with leave being more common among women and people of color of all genders than their White male counterparts (Armenia & Gerstel, 2006; Herr et al., 2020). For this reason, leave policies should make clear that caregiving leave cannot be used for research activities, and to tie leave eligibility (along with tenure probation extensions) to birth recovery or caregiving of at least 20 hours per week (Burch et al., 2023; Williams & Lee, 2016). To the committee's knowledge, however, the efficacy of these various leave policy options has not been assessed empirically.

Another approach to address the unique needs of birthing employees is to layer forms of leave. While all caregivers should be permitted to take time off when needed to provide care, birthing parents need to care for infants and simultaneously address their own health needs. As such, it would be appropriate to allow these employees to take additional time specific to their health. The best practice is to provide all caregivers a 12-week leave, and birthing caregivers could additionally be entitled to a leave term to account for their temporary incapacity (Williams & Lee, 2016). This could be covered under an institution's disability leave policy to supplement the caregiving leave for birthing parents.

Research has found that it is more effective for leave to be provided as a standard benefit for employees rather than establishing onerous application requirements that may limit uptake (Roselin et al., 2022). Length of leave should also be made clear. Twelve weeks is generally seen as a standard for leave length and typically works well for employee leave policies. Universities, however, also need to consider other populations. Twelve weeks may work well for faculty and staff, but some research has suggested it can present challenges for student populations. Daskalska et al. (2022) note that lengthier leave terms can affect students' ability to graduate on time. The authors of this study suggest institutions start with a 6-week leave policy for students with additional coursework accommodations to balance the need for leave, with the desire to maintain academic progress. Ultimately, more research is needed to determine the length of leave that is most beneficial to students beyond a 6-week minimum.

For students, taking leave may result in financial costs through accrued student loans and delayed earnings. Students may also lose access to health benefits, financial aid, and other benefits and support of enrollment while on leave (Kraus et al., 2021; Roselin et al., 2022). To address this, Roselin et al. (2022) found that several undergraduate medical programs ensured their students retained access to campus supports by providing an "enrolled academic adjustment" option allowing their students to reduce their academic duties while preserving their student status. Other adjustments could

BOX 6-4
Examples in Action: Retaining Clinical Scientists

To help retain clinical scientists, the Doris Duke Charitable Foundation provides awards to medical schools to offer extra research support for faculty who face substantial caregiving demands in order to support continued research productivity of those with caregiving responsibilities as well as to increase awareness of the need for support (Jagsi et al., 2022). Examples of extra support include additional laboratory staff and help with grant and manuscript writing. The program expanded during the COVID-19 pandemic with additional funders joining the effort and a focus on supporting women of color conducting biomedical research. As part of the program, the foundation also enlisted scholars to evaluate its impact. Evaluation of the program suggests an important effect of the program in transforming culture away from stigmatization and toward validation of caregiving responsibilities while directly addressing specific needs of faculty with caregiving responsibilities (Szczygiel et al., 2021).

BOX 6-5
Best Practices for Policies Related to Accommodations and Adjustments Checklist

- ☐ Provide accommodations for faculty, staff, and students not only in situations where it is legally required but wherever necessary for caregiving responsibilities. This is desirable for those individuals and will help eliminate stigma for those with caregiving responsibilities, too.
- ☐ Ensure that faculty members have a central, trained point of contact in the human resources department, the provost's office, or the diversity, equity, and inclusion office who will help them negotiate pregnancy accommodations.
- ☐ Provide stop-the-clock policies for anyone with a demonstrated need (not just for mothers, or parents) who certifies that the time will be spent on caregiving, to clarify that this is not a free research leave.
- ☐ Train faculty not to penalize those who stop the clock.
- ☐ Provide standard language in all rank and tenure letters to ensure that those who have stopped the clock are considered as having years-in-service that omits the stop-the-clock period.

include reducing course loads, allowing for virtual participation in coursework, and providing evening courses and accelerated programs (Clark et al., 2021; Contreras-Mendez & Cruse, 2021; Huerta et al., 2022; Ryan et al., 2021; Springer et al., 2009; Wladis et al., 2023).

Resident physicians also encounter challenges in taking leave and the ways in which this may delay progress and interrupt timelines in their programs. The American Academy of Family Physicians (AAFP) encourages programs to consider "home-study or reading electives" to minimize the time residents need to take off for family caregiving (Weinstein et al., 2019). AAFP also suggests proactively planning rotations to minimize disruption around the expected birth or adoption of a child and ensuring there is no "on-call time" during leave. Importantly, this requires planning to ensure the peers of residents on leave will not be required to make up the work. Programs can do so by including deliberate redundancy in staffing plans, or funding coverage by external staff (Weinstein et al., 2019). Part-time options may be useful for other medical trainees as well. For example, Weinstein et al. (2019) have called for the initiation of part-time graduate medical education, or GME, tracks, which could allow trainee physician caregivers to continue their education while meeting their family's needs.

For faculty, stop-the-clock (STC) policies and other duty modifications can be important. It has now been nearly two decades since a report

---

BOX 6-6
Examples in Action: Supporting Postdoctoral Caregivers

Following organizing efforts, in the 2022 contract between the University of California system and postdoctoral scholars, postdocs won 8 weeks of fully paid leave that can be used after the birth or adoption of a child, or for family care. Postdocs can also use paid time off, sick leave, and short-term disability for maternity leave. Childcare subsidies[a] of $2,500/year, increasing to $2,800/year by 2026, and lactation support at work (including access to private space and time for breastfeeding) are also included in the contract. The subsidies apply to childcare costs for qualified dependents age 12 or under who reside with the covered postdoc.

---

[a] While these subsidies can provide important assistance, it should also be acknowledged that the average cost of childcare in California is estimated at over $20,000 a year (Cutler, 2023).

from the American Council on Education strongly recommended that any penalties in the hiring or tenure process resulting from caregiving gaps be abolished (Marcus, 2007). Among other things, the report advised universities to allow faculty to extend the tenure probationary period by up to 2 years following a child's birth or adoption. Many institutions already meet these goals. A major study of family-friendly academic policies conducted in 2007 concluded that "one of the biggest problems ... isn't that these kinds of benefits are not available for faculty with families. It's getting faculty to take advantage of them" (Marcus, 2007). Both men and women express concern that they will be judged harshly during the tenure process if they have stopped the clock or taken family leave (Moors et al., 2022; Sallee, 2008).

As noted in Chapter 4, however, several studies have documented gendered consequences of STC policies of promotion and tenure. While these issues deserve more evaluation, experts have proffered several best practices designed to reduce the likelihood faculty feel they need to ensure their extension period remains "productive" and ideally reduce discrepancies for faculty who need to utilize a tenure extension period to address other needs. As noted before with caregiving leave, policies that normalize and require using leave for caregiving and not further advancement of research are important. Specifically, for STC policies, one solution is to implement agreements that faculty cannot use research products started during the leave period in tenure portfolios (Burch et al., 2023).

Another suggestion for reducing stigma around STC policies is for them to be opt out rather than opt in. Such policies normalize their use, sending the message that usage is expected (Burch et al., 2023). In 2005, Princeton changed its STC policy to require faculty to opt out, because so few faculty had sought to use the policy when it was structured on an opt-in basis. Women surveyed as part of the study that led to the policy change reported concern about being seen as less focused and less committed to their work should they elect to stop the clock (Marcus, 2007). And opt-out policies have been shown to produce benefits for the representation of women and particularly women of color among tenured faculty (Gonsalves et al., 2022).

An important element of an effective stop-the-clock policy is a requirement that notifies appointment and promotions committees that a faculty member who has stopped the clock should be evaluated as if they had the same number of years in service, not as if they had been in service during the period when the clock was stopped (Antecol et al., 2018; Beckerle et al., 2011; Ecklund & Lincoln, 2016). Unless the relevant committees receive this guidance, STC policies can end up penalizing those they are designed to help.

Along with STC policies, there are various ways to structure and modify duties that could be used to accommodate the different needs and experiences of faculty performing different kinds of caregiving duties (for more details on duty modifications, see Chapter 7). Possible interventions include part-time work, job sharing, and adjusted tenure timelines. All these policies should be framed as career enhancing rather than limiting (Ibrahim et al., 2017).

Compared with part-time work in other sectors, specific challenges can arise when orchestrating part-time faculty appointments. With faculty work split across the functions of teaching, research, writing, and service, it is not always obvious which components of work can be cut back. Research supports offering faculty half-time appointments at full benefits, as well as temporary reductions of full-time appointments on a short-term basis to support faculty facing a particular life event (Koppes Bryan & Wilson, 2015). Studies of work-life support for academics also reference job sharing (or dividing one faculty position with benefits into two part-time positions with benefits) as another option (Koppes Bryan & Wilson, 2015) (for more detail on job sharing, see Chapter 7). Although there are precedents for job sharing, particularly internationally (Stoller, 2023), little research has tracked the prevalence of such policies and efficacy in the U.S. context (Koppes Bryan & Wilson, 2015). Importantly, these part-time appointments should not be marginalized and considered contingent gig work within the academy; instead, they should be treated as a valued, alternative path.

Most literature about modifications and accommodations relates to pregnant academics and new parents; additional information is needed regarding the interventions most helpful to employees who need modifications due to caregiving for family members and loved ones other than babies and children. For example, a 2007 study found that faculty providing older adult care were less likely to want to reduce their work hours than those caring for children (Keene & Prokos, 2007). Caregivers for adults may also need different communications approaches. A University of California, Davis, study of clinical and/or research biomedical faculty found that 24 percent of male and 14 percent of female faculty respondents needed accommodations to care for an adult or due to the death of a close family member (Shauman et al., 2018). Many of the existing family-responsive policies were not well utilized by these employees, as they were overwhelmingly unaware of benefits or held misconceptions about the programs. Participants also reported that supervisors served as gatekeepers to policies—even in cases that they were not required to approve a particular benefit. Shauman et al. (2018) suggest this points to the need to train department chairs and other supervisors so that they are aware of policy details and the importance of facilitating access to the policies.

Policies Related to Direct Care Support

Caregivers in academia experience significant financial strain caused by both caregiving (e.g., paying for childcare) and its relationship to work (e.g., salary losses from taking leave or delaying tenure) and/or education (e.g., enrollment, persistence, and graduation). When caregivers are unable to meet their basic needs, their work and career suffer. For students, this is especially relevant. While basic needs insecurity is prevalent for most college students, the rates are especially high among parenting students, the majority of whom are students of color (Institute for Women's Policy Research & Aspen Institute, 2019). In a 2019 survey, the Hope Center for College, Community, and Justice found that among 23,000 parenting students, 53 percent were food insecure in the last 30 days, and in the previous year, 68 percent had been housing insecure, and 17 percent have been homeless (Goldrick-Rab et al., 2020), and these inequities were only further exacerbated by the COVID-19 pandemic (White & Cruse, 2021). These kinds of inequalities are also important to note because while most individuals will care for a family member during their lives, women, LGBTQ+ individuals, and people of color are more likely to be driven into poverty as a result of this (Bose et al., 2021).

Given the challenges outlined above, caregivers need to have their basic needs met to most effectively perform their work or engage in their education. Centralizing support and providing resource navigation is an important practice for ensuring caregiving students are aware of and can most efficiently access basic needs programming organized in spaces that are often called student parent resource centers and provide information on childcare, grants, family housing, student and dependent insurance, and more (Coronel, 2020; Goldrick-Rab et al., 2020; Mason, 2022; Mason et al., 2007; Springer et al., 2009). Creating a university position specific to supporting student parents and their families can also help to institutionalize and better ensure the continuation of policies aimed to help family caregivers (Robertson & Weiner, 2013). Student parent survey participants have noted that "holistic

> **BOX 6-7**
> **Best Practices for Policies Related to Direct Care Support Checklist**
>
> ☐ Provide centralized access to needed resources such as childcare, insurance, and housing support in a single resource center or office.
> ☐ Institute a family resources officer who can serve as a point person for care support and other related needs.
> ☐ Develop on-site, affordable child and adult care options for faculty, staff, students, and trainees.
> ☐ Offer subsidies to support accessing quality care.
> ☐ Offer care-related travel grants to support travel for care recipients at conferences.
> ☐ Proactively educate workers on adult care benefits and resources.

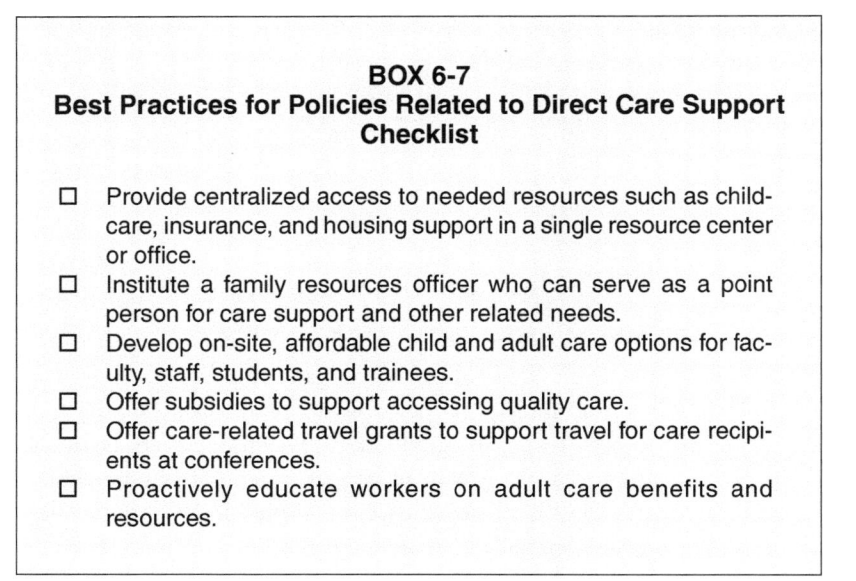

wraparound support," including addressing their basic needs, was critical in ensuring their success (Contreras-Mendez & Cruse, 2021). Such efforts can be transformative. In a case study from a large mid-Atlantic university during the 2009–2010 school year, a campuswide movement had been initiated to collect a small student fee for resources for caregiving students. The fees collected created a subsidy for childcare, a dormitory for families, and lactation rooms on campus resulting in a 93 percent retention rate for pregnant and parenting students (Brown & Nichols, 2013).

> **BOX 6-8**
> **Example in Action: Backup Care**
>
> The University of Maryland provides all faculty, staff, contingent employees, and graduate assistants with access to the Care@Work platform which connects individuals with paid caregivers. The university covers the cost of monthly membership and subsidizes up to 10 days of backup care in instances when regular care arrangements are not available. Similarly, the Icahn School of Medicine at Mount Sinai subsidizes 10 days of emergency backup care with Bright Horizons for all staff, trainees, and faculty, with fees scaled to income.

Childcare

For students, postdocs and other trainees, faculty, and staff alike, access to different forms of paid caregiving is also crucial. While already well established, the COVID-19 pandemic dramatized the importance of access to childcare for academic parents (Bender et al., 2022). The vast majority of research on best practices to support caregivers in higher education proposes that institutions develop and maintain on-site childcare options that offer proximity to and an alignment with parents' work needs (Cardel et al., 2020; Carr et al., 2017; Ibrahim et al., 2017; Ladores et al., 2019). In addition to meeting the basic needs of working parents, on-site childcare can signal a climate that is broadly supportive for caregivers (Carr et al., 2017). At the same time, universities should be thoughtful that this approach may not be preferred by all members of the institution. In particular, some research shows that Black mothers prefer to utilize other forms of childcare, such as kin-care or community care centers, given both concerns of potential racism in predominantly White care settings and a reflection of cultural preferences (Dow, 2015; Uttal, 1996). Given this, options for care subsidies or reimbursements for different arrangements off campus are additionally important.

BOX 6-9
Example in Action: Supporting Pregnant and Parenting Students

Sacramento State University offers a Parents and Families Program within its Division of Student Affairs to provide support for pregnant and parenting students at the university. The website provides easily accessible links to the various programs and resources available to students with children. For pregnant students, the university has a pregnant student liaison, who works with them one-on-one to develop a plan for their birth and return to school and for contacting their professors as well as for what to do if this plan needs to change. For students with children, the university provides federally supported childcare on campus, information on resources across the university, and access to student parent ambassadors, who are current students with children who have volunteered to help the community, provide support, and advocate for the needs of student parents. The Parents and Families Program home page also links to a map for the locations of all diaper-changing stations and mother's rooms on campus.

These recommendations are in line with the unique time demands of a career in STEMM, which are less likely to be limited to the business hours of day care centers. Affordable on-site childcare, while largely beneficial to academic caregivers and the institutions that employ them, is just one of a wide array of strategies for meeting parents' childcare needs and may be a better fit in some institutional contexts than others. A comprehensive 2020 study of best practices for supporting women and caregivers in STEM at universities in the United States recommends the development of a comprehensive suite of childcare support (Cardel et al., 2020). This includes a university childcare and family resources web page, and a point person who can serve as a family resources officer. Universities should also offer backup and emergency care options, childcare for snow days, childcare for any public-school holidays that conflict with university schedules, and access to summer camps nearby. They should also subsidize the costs of work-related travel to offset childcare expenses (Cardel et al., 2020). As noted in Chapter 5, however, the provision of childcare resources is not enough if the diverse needs of neurodivergent children and children with disabilities are not accounted for and parents of these children cannot access the more specialized services they need. In designing childcare resources, administrators need to ensure they are not implicitly assuming they are only providing support in the care of neurotypical children without disabilities.

As noted in Chapter 5, universities also need to be mindful of challenges of availability. Certainly, increasing childcare options on campus can help, but creative solutions may also be useful if a broad increase in on-site care is not possible. Such solutions could include allowing families who do not require full-time care Monday through Friday to select the days and times that they require, allowing for flexibility such that another family could utilize open slots. Additionally, universities can and should apply for federal and state grants such as the Child Care Access Means Parents in School Program, or CCAMPIS, which provides funds to support or establish childcare centers on campus to serve low-income students (U.S. Department of Education, 2023).

A large body of research additionally highlights the challenges that faculty, graduate students, and postdocs and other trainees experience with securing childcare during conference and fieldwork travel. These challenges tend to disproportionately disadvantage women, who continue to shoulder more caregiving responsibilities in families, and contribute to the "leaky pipeline." Travel for research and conferences facilitates high-quality published work and is essential for networking (Knoll et al., 2019). It is, in other words, critical for career development, particularly for junior scholars (Tower & Latimer, 2016). Some professional organizations, such as the American Society for Cell Biology, provide travel awards that parents can use for any combination of childcare support that will allow them to attend the annual meeting and present their science, for example, extended childcare hours, transportation, and accommodations for children and a caregiver along with the scientist parent, and transport of a relative to the scientist's home to care for children in the parent's absence (American Society for Cell Biology, 2022).

Universities, professional associations, research societies, and individual conferences all have a potential role to play in easing the burden on caregivers for their work travel. Recent scholarship on the challenges surrounding work travel for caregivers in higher education strongly recommends that universities and other organizations do more to offset faculty's caregiving burden—whether for childcare, older adult care, or care for adult dependents with special needs (Baldiga et al., 2018; Calisi & Working Group of Mothers in Science, 2018; Tower & Latimer, 2016). There are different ways to support childcare either at home or at conferences and events. The recommended best practice is to offer faculty a variety of options, including financial support that allows families to meet their own needs and generously reimbursing reasonable expenses on a case-by-case basis.

The needs of caregivers vary widely based on the age and needs of their dependents. For example, a lactating parent might prefer financial support to travel with their baby and a partner or other caregiver, rather than receiving a subsidy to pay for extended childcare at home. The parent of an older child might need an arrangement that allows their child to remain at home and attend school. For parents caring for neurodivergent children or children with disabilities, needs may be even more complex and multifaceted. Depending on the child's specific needs, parents may require assistance from a companion when traveling or need support for engaging a family member or trusted paid caregiver to look after their children while they are away. Regarding work travel, those with caregiving responsibilities in different disciplines may have different needs, even within STEMM. For instance, fields with field-based research obligations such as geology may pose significant obstacles for breastfeeding parents. Even within universities, departments should evaluate the unique challenges faced by their faculty caregivers to offer the most appropriate forms of support (Baldiga et al., 2018). All organizations involved in supporting caregivers traveling for work should assess and improve available options on an ongoing basis to ensure that they are meeting family's needs (Boss et al., 2017).

Older Adult and Adult Dependent Care

As stated previously, nearly 20 percent of U.S. adults are providing unpaid family care to an adult age 18 and older (AARP & National Alliance for Caregiving, 2020). However, the needs and solutions for providing older adult care at academic institutions have been less studied than childcare. A recent article examining older adult care in academic medicine highlighted not only the general lack of investment in older adult care in the United States but also the ways in which current policies at many institutions are not designed with the unique needs of those caring for adults in mind (Sosa & Mangurian, 2023).

Best practices for providing older adult care and adult care support are myriad, yet they tend to center on assisting employees with securing care and reducing the mental strain of managing caregiving and end-of-life planning (Koppes Bryan & Wilson, 2015). For example, institutions have offered a variety of supportive programming, from support groups to respite care and meal preparation; however, the efficacy of these popular interventions is largely unstudied (Skarupski et al., 2021). Among early-career faculty surveyed at one university medical center, the most preferred types of caregiving assistance were the provision of a laboratory technician at work, a personal assistant or coach, and general household help (Hartmann et al., 2018). Preference for these forms of assistance varied by gender, with the

---

BOX 6-10
Best Practices for Caring for Adult Dependents Checklist

☐ Federal law prohibits discrimination against anyone caring for an individual with a disability, whether that individual is a family member or not.
☐ Identify students caring for family members other than children and make similar supports available to them.
☐ Partner with university departments focused on aging and geriatric medicine to develop centralized resources to help guide those providing support to older adult dependents on the policies and practices available to them.
☐ Allow for caregiving leave policies to be employed incrementally throughout the year to address the needs of those caring for older adults who may need time off in regular increments for doctors' visits or other needs.

most popular intervention among women being household help and among men, the assistance of a laboratory technician at work.

In contrast to expectant parents, those caring for adults are less likely to have their caregiving status or needs known in the workplace (Dembe & Partridge, 2011; Gabriel et al., 2023). As such, it is particularly important for employers to proactively educate workers on their older adult care services and benefits (Calvano, 2013). Underutilization and fear of disclosure are significant barriers to employees gaining the full benefits of the programs (Calvano, 2013; Dembe & Partridge, 2011). In a study of faculty at a large research institution, 91 percent indicated they were unaware of their university's policies and procedures for older adult care (Leibnitz & Morrison, 2015). Many reported difficulties finding the information they needed, with STEM faculty being significantly less likely to know institutional policies. As such, institutions should consider tailoring their offerings and communications strategies so that they are more effective in reaching those in need.

Important in the considerations of older adult care is the life stage of those faculty and employees most likely to need it. Though caregiving of older adults can occur at any stage, the "biggest squeezes" occur in early adulthood (primarily with children) and in the years preceding retirement (Patterson & Margolis, 2019). As such, institutions may want to consider offering older adult care support as a critical tool particularly for retaining

---

BOX 6-11
Example in Action: Supporting Graduate Student and Trainee Caregivers

The University of California, San Francisco, provides a centralized hub through their MyFamily portal, which provides links to resources for all members of the campus community with caregiving responsibilities. For graduate student workers, the university provides up to 10 weeks of paid parental leave following birth, adoption, or placement of a foster child. Ph.D. students are also entitled to up to 4 weeks of leave to manage their own health needs or to care for a qualifying family member. The university also provides access to resources to help protect parenting graduate student employees working in labs as well as a list of financial resources for graduate students with dependents.

later-career faculty. There are limited data on the ramifications of older adult care support and retirement decisions in STEMM broadly, but studies in the medical field are instructive. Women physicians have fewer challenges with work-life integration in senior years, but they are still prevalent (Templeton et al., 2019). Faculty women in academic medicine were more likely than faculty men to be caregivers and to cite caregiving and health care as important factors in one study of retirement decisions (Levine et al., 2022). In contrast, another study found no significant difference in retirement intention related to gender and caregiving; however, this may have been influenced by the relatively young ages of the faculty in the sample (Skarupski et al., 2021). It may be that later-career faculty women are choosing to delay retirement due to caregiving demands; a study across several institutions found that 51 percent of late-career women faculty reported the need to care for relatives as a reason to delay retirement, as compared with 37 percent of men (Berberet et al., 2005). More research is needed to understand this dynamic and the potential to address gender and caregiving gaps in the late-career professoriate. Addressing the older adult care crisis may be particularly important in light of the benefits of faculty mentorship by senior women and people of color.

There is no national directed effort geared toward supporting mid- or later-career faculty with caregiving demands, unlike early-career faculty, who are more often managing the demands of childrearing (Skarupski et al., 2021). Institutions should ensure that, at minimum, their existing support policies (e.g., sick leave, flextime, altered working schedules) are marketed toward those who might need older adult care support, not merely those who are parenting (Leibnitz & Morrison, 2015).

BOX 6-12
Example in Action: Adult Care

Virginia Commonwealth University runs a Family Care Center on campus that is accessible to all health employees at the university. The center provides both child and adult day care on campus and is run through their Family Centered Programs department. This center was the first of its kind in the region.

CULTURAL SHIFTS TO CHALLENGE IDEAL WORKER NORMS

All the practices mentioned in this chapter can help to support family caregivers in academic STEMM, and the adoption and dissemination of these policies and practices can also serve to inform and express the priorities and expectations of institutions. Yet, even with the best policies and practices in place, the cultural norms in academic STEMM continue to limit utilization. In fact, recent research reported in *Nature* reveals that a toxic workplace culture is the main reason women leave academia (Sidik, 2023).

Work-life initiatives remain on the peripheries of organizational discourses and strategies, rather than as core dimensions of academic culture (Ernst Kossek et al., 2010; Kossek & Lee, 2022; Valantine & Sandborg, 2013), contributing to a culture that discriminates against caregiving. As Kossek et al. (2010) write, "work-life changes ... [need to be] part of the core employment systems to enhance organizational effectiveness and not just as strategies to support disadvantaged, non-ideal workers." Rather than an addendum, work-life inclusion must be foundational to the creation of university policies, to help dismantle cultural models of work and redefine academic excellence and success with flexibility in mind (Blair-Loy & Cech, 2022; Ernst Kossek et al., 2010). Without transparency and accountability, cultural schemas (i.e., the status quo) prevail, along with the accompanying biases that negatively affect the health and well-being of individuals and the state of scholarship and innovation more broadly (Christensen, 2013; Valantine, 2020).

To situate caregiving issues as central to the organizational culture of higher education, scholars and university administrators can adopt a range of strategies—at individual, departmental, and institutional levels. Offering a simple, individual-level strategy, Arora et al. (2020) propose the "COVID-19 CV Matrix" as a potential framework for documenting contributions, disruptions, and caregiving responsibilities during the pandemic, to aid in fair evaluations from tenure and promotion committees. Specifically, a sample matrix might consist of three columns, listing categories (e.g., research, education, media), activity (e.g., halted, prep for online transition, op-eds), and descriptions (e.g., reported to organization, summer research course, *New York Times*), respectively. Seeking to address concerns that the initial matrix did not adequately address the gendered and racialized impacts of the pandemic, Raja et al. (2021) have created a "CovidCV prototypical system," which "creates a color-coded CV from the user's data entries documenting work and home life." Specifically, users can document academic successes and setbacks, family events, such as birthdays, and ongoing struggles, such as the caring for a loved one, marking each entry as major or minor, good, bad, or neutral, and the feelings or emotions over the course of a given week (Raja et al., 2021). The goal of this system is to provide "the underlying 'invisible context,'" by illuminating the conditions, events, and struggles that affect each faculty member's ability to work and live in a holistic way (Raja et al., 2021). While created in the context of the COVID-19 pandemic, such a system could have continued benefits as it allows for the acknowledgment of life outside of work.

In addition to individual-level efforts, meso-level interventions, at the everyday level, are needed to provide educational and workplace supports for fostering equity and inclusion in academic environments (Kossek & Lee, 2022; Kossek et al., 2011; Ward & Wolf-Wendel, 2012). For example, Kossek and Lee (2022) examine how work-life issues intersect with gender and affect women's career advancement (focusing on the business school context) and recommend actionable steps for leaders. Specifically, they highlight the importance of "work-life boundary control," or the ability to control the separation, integration, and salience of work and nonwork roles to avoid role conflict or strain. Actions that can aid in this boundary management include scheduling meetings and events with family responsibilities in mind (i.e., from 9 a.m. to 3 p.m.); encouraging email breaks and vacations; celebrating egalitarian caregiving efforts; and recognizing nonwork achievements (Kossek & Lee, 2022). Notably, these suggestions are feasible and can be enacted in programs, departments, and colleges to target academic organizational cultures by placing work-life inclusion at the forefront.

At the institutional level, cultural solutions can also be implemented, especially through efforts to change leadership to engender cultural change (Valantine, 2020; Valantine & Sandborg, 2013). In fact, Valantine and Sandborg (2013) called for 50/50 leadership representation among women and men in academic medicine by 2020, arguing that closing the gender gap in leadership would help to usher in improvements in work-life integration and flexible work options. Though gender parity in academic leadership in science and medicine has yet to be achieved, Valantine (2020) notes its continued importance as a "seed [for] the cultural change necessary for inclusive excellence" and documents the system-level strategies that have been implemented in the National Institutes of Health's (NIH's) intramural research program to foster an organizational culture of inclusion and equity.

The NIH's four integrative strategies include having an Equity Committee to track metrics of diversity and inclusion; diversifying candidate searches beyond personal, informal networks; providing bias educational training for all search and promotion committees; and establishing the Distinguished Scholars Program of roughly 15 tenure-track investigators with commitments to inclusive excellence (Valantine, 2020). Uniting the components of this plan is the common goal of changing the institutional culture of academic medicine through the increased representation of women and other underrepresented groups. Such strategies for cultural change can be used as a model for other institutions and will also lend support to individual actions (such as CV framing), as well as departmental or college-level efforts (such as work-life boundary management).

SUMMARY OF FINDINGS FROM CHAPTER 6

The challenges family caregivers currently face in academic STEMM are significant, and prevailing cultural norms and schemas can create significant barriers to addressing these challenges. No one solution provides the silver bullet that ensures a more welcoming and inclusive environment for family caregivers with the flexibility they need because the needs of family caregivers are diverse and complex. Instead, many opportunities exist to implement an array of policies and practices that can provide broad and flexible support to family caregivers in academic STEMM whether they are students, trainees, faculty, or staff. These best practices are not only nice to have, but highly beneficial to promoting an inclusive and welcoming STEMM environment that helps to further build and support this workforce and its innovation.

1. Ensuring legal compliance with requirements established by Title IX, FMLA (federal, state, and municipal), the Pregnant Workers Fairness Act, Title VII, and other key provisions established to protect caregivers is a necessary starting point for supporting caregivers in academic STEMM.
2. The legal patchwork is so complex that it is impractical to expect individual students or faculty to figure out the relevant legal rights and obligations that apply to them. An important best practice minimum is that Title IX offices be fully trained in students' legal rights, and that students know when they should consult with Title IX officers. Equally important, leave and accommodations

decisions should not be left up to individual faculty members or department chairs without guidance from someone who is expert in navigating the complex legal environment, preferably a specially trained person in human resources with additional backup by someone in the provost's office for faculty.
3. Foundational best practices for providing caregiving leave include adopting a formal policy with clear standards across the institution, ensuring gender-inclusive access to leave, ensuring leave is not used for research activities, offering leave as a standard benefit rather than requiring an onerous application process, providing paid leave, and making policies promoting flexibility opt out rather than opt in.
4. Many caregivers require accommodations to allow them to perform their roles both as caregivers and as staff, faculty, or students. Accommodations include course adjustments for students; part-time options for faculty, staff, and trainees; and stop-the-clock policies for faculty on the tenure track.
5. Bias and discrimination against caregivers still exist in STEMM. To address this, best practices include providing pertinent information in basic antibias trainings and in specialized trainings for search committees, department chairs, and the like.
6. Caregivers can face significant financial strain and would benefit from supports that address their basic needs, such as on-site childcare or childcare subsidies, support for adult care, and providing easy access to available resources online.
7. For policies supporting caregivers to have the biggest effect, cultural change is needed. Though challenging, important drivers of culture change exist at individual, departmental, and institutional levels to build greater trust, transparency, asking and listening, collaboration, and accountability.

7

Innovative Approaches to Career Flexibility

This report thus far has presented established approaches that have been tested and improved over time for best practices that support caregivers through better and more accessible information about resources, more control over weekly work schedules, and stopping the tenure clock and modifying duties. But, as noted in Chapter 6, cultural barriers remain and inadequate support for caregivers still exists, with repercussions for retention and advancement for women and other family caregivers. Action cannot only look to what has already proved to be beneficial but also should push beyond current barriers and strive for even greater support for family caregivers. This chapter details innovative practices as a call to action for new and inventive solutions that encourage creative thinking.

A common theme that has emerged is the importance of flexibility and experimentation to successfully maneuver competing responsibilities. Rigidly organized traditional science, technology, engineering, mathematics, and medicine (STEMM) workplaces do not serve our workforce and do diminish its ability to perform in the creative and scholarly ways for which they have trained. Data from Future Forum find that flexibility is just behind compensation in terms of factors affecting employee satisfaction, and the vast majority of workers want flexibility in both where and when they work (Future Forum, 2022). Moreover, many newer entrants to the workforce expect flexibility and support for personal and family life as a critical job support (Friedman, 2022).

In response to these challenges, the committee sought out workplace examples of innovative ideas, ones that go beyond currently established policies to tackle cultural change and provide greater support to caregivers particularly in the realm of flexibility. There is also the need for innovative solutions to support students, and thinking creatively, the examples presented here could be adapted to their needs.

The following innovative approaches do not have the same degree of research backing and evaluation as the best practices outlined in Chapter 6, but they encourage us to look beyond current practices to consider what the future may hold for supporting family caregivers. The committee also draws on the words of family caregivers themselves (see Boxes 7-1 and 7-2) to detail the kinds of actions they see as beneficial to reimagining the academy to be more supportive of family caregivers.

As with the best practices detailed in Chapter 6, the potential for negative and unintended consequences always remains, even more so when policies are new and innovative. Engaging innovation remains important to advance support for caregivers, but care must also be taken to ensure attention to how policies are implemented and ensure adequate evaluation of both success and challenges to adapt and build further.

From the sample programs and practices covered in this chapter, the committee suggests five drivers of good culture that stand out and could

---

BOX 7-1
Reimagined Academy: Alternative Visions of Academic Success

Caregivers put revised standards and process for tenure and post-tenure advancement at the center of their alternative visions for the academic STEMM workplace.

Building on the idea of a "COVID statement" (implemented on some campuses for faculty who came up for advancement during the early pandemic), interviewees suggested institutionalizing a process for taking ongoing caregiving responsibilities into account in academic advancement decisions. Several interviewees proposed that all tenure and promotion packages should include a written statement on responsibilities outside of the university.

serve as effective antidotes to a toxic culture: trust, transparency, asking and listening, collaboration, and accountability. Thinking creatively about each of these drivers can spur action toward new and unique ideas.

Trust: Current forms of management typically do not engender trust in employees. Focusing on how many hours a person sits at their desk is not a measure of their productivity. Trust is evident when employees are empowered to set a schedule for themselves that meets their needs, as well as the needs of their employer.

Transparency: Transparency involves the quality of being open to public scrutiny. In so many workplaces, private deals abound—when you work, where you work, or how you work, as well as your compensation. Transparency about these arrangements and requirements for advancement and tenure increases the possibility of equity plus the recognition of individual needs and responsibilities.

Asking and listening: Many workplace innovations in both scheduling and work processes do not come from top-down fiats. They come from asking employees what works, what does not work, and what they need, and then listening to their answers. Employees are closest to the work and often know better than others what needs to change to support their needs and promote their productivity.

A packet for promotion, for tenure, for evaluation would contain a section that describes your outside responsibilities, your home responsibilities.... 'Tell us about who you are outside of work and what other responsibilities you have.'

Interviewees envisioned a caregiving statement detailing current (and relevant recent past) caregiving responsibilities, and perhaps other outside activities that conferred a sense of the candidate as a person with compelling investments and accomplishments beyond the academic workplace.

"The timeline maybe wouldn't need to be so rigid, or even the expectation of actual amounts of productivity, if you could take the opportunity to explain: "This is what I did with my time. Despite the fact that I have X, Y, Z other responsibilities to take care of at home ... I was still able to manage to do this thing in this amount of time with what I have."

> **BOX 7-2**
> **Reimagined Academy: Care-Centered Academic Workplace Norms**
>
> Many interviewees spoke to the need for radical change in the culture of academic STEMM workplaces in the United States, rejecting what they saw as the primacy of competition and individualism over care and collaboration.
>
> Caregivers of color and those from immigrant backgrounds in particular made a compelling case that academic STEMM workplaces would do well to adapt some of the assets reflected in non-dominant cultures, particularly around valuing care and collaboration.
>
> "When you become a caregiver, you acquire so many skills that are very helpful in teamwork. So, I think if academia starts valuing more than achievement, how good you are as a collaborator, I think a lot of caregivers coming from ... a minority background will be very successful because by definition ... it's not by definition, by history, I mean the people who come from minority backgrounds, they have to struggle. So, there are a lot of valuable skills in that journey that can be very helpful in teamwork."
>
> Interviewees advocated incorporating "the values of caregiving" into the broader culture of STEMM workplaces. Rather than making special accommodations for caregivers, some suggested universal changes that made academic STEMM workplaces more consistent with a balanced life and substantial priorities outside of work.

Collaboration: Today's work environments often center on teams. We see this in the private sector as well as in academia in shared jobs and shared laboratories, as described below.

Accountability: Any workplace change demands accountability. Is it reaching the goals it set? Are the individuals being held accountable? Accountability is built into several of the examples provided below.

Each of these drivers plays an important role in exploring novel approaches for university supports for family caregivers and allows individuals to be their most authentic selves. The committee presents them next to ignite imagination about how organizations can rethink work within a framework that is compatible with workers who are caregivers.

> "We have a rule here that ... everything has to be done by 5:00 because daycare pickup is 5:30. So, as a department, we don't do anything that starts after 4:00. It's a very simple policy, but it's very impactful."
>
> Universalizing a culture of work-life balance, they suggested, could place caregivers on more level footing with their peers ("nobody's not showing up, nobody's not doing their job"). But it also promoted work-life balance for everyone in the department. Many caregivers made the point that an academic STEMM culture that valued caregiving would be healthier for non-caregivers as well.
>
> "If caregiving values were incorporated, I think standards would just be a little bit more realistic, and I think there would potentially be more room for people to be a little, a little more sane. Just because there would be room for that, like, life balance. Because, like, I said, I don't think just because if your kids are older, or if you don't have kids, I don't think there's any harm in taking breaks. I don't think caregivers exclusively need that. I think *people* need that."
>
> Interviewees suggested that valuing care and recognizing caregiving responsibilities was central to cultivating a more holistic, less mechanistic regard for oneself and others in the STEMM workplace. As one interviewee who had supervisory responsibilities explained,
>
> *"It's [about] preparing for contingencies ... having awareness that we're working with people, we're not working with machines, and that people have needs."*

INNOVATIVE APPROACHES TO PROMOTE WORKPLACE FLEXIBILITY

Temporary Decreased Effort

Disruption of on-site work patterns in academic institutions caused by the COVID-19 pandemic led many to look for ways to incorporate the potential for temporarily decreased effort and to explore alternative work options. Several universities have previously led and currently lead the way in addressing the need for transparent and accountable guidelines for negotiating part-time faculty positions.

Ten years ago, UMass Chan Medical School established criteria and a process for requesting part-time status.[1] "These guidelines provided a clear and transparent framework to be used by both faculty and chairs in the discussion and subsequent decision about proposed reduction from full-time to part-time effort. The guidelines also clarify the process for decision-making and monitoring" (Thorndyke et al., 2017). The guidelines were implemented to ensure greater consistency in part-time work across the UMass health care system. A survey of department chairs and chiefs after implementation found that one-third had made use of them to discuss part-time options with faculty and found the guidelines to help in determining whether part-time work was an option and what approach made sense (Thorndyke et al., 2017). One important guideline allowed faculty initially hired into full-time positions to temporarily negotiate part-time and then return to full-time work. Similarly, in the University of California system, academic appointees may be eligible to reduce their percentage of time of an appointment from full-time to part-time for a specified period or permanently to accommodate family needs.[2]

Another version of flexibility is the Massachusetts Institute of Technology initiative, Work Succeeding, which provides rigorous guidelines as well as toolkits for managers and employees primarily focused currently on a variety of work sites: in-person, remote, and hybrid work rather than on decreased effort.[3]

University and medical school guidelines that emphasize transparency and accountability provide structure for both leadership and faculty in navigating the process to allow for part-time or off-site positions.

Teams

The significance of teams to higher education is increasingly apparent. Teams with scientific, educational, or administrative missions provide benefits that include creative solutions from differing perspectives. Collaborative

---

[1] To learn more about the UMass Chan Medical School part-time guidelines, see https://www.umassmed.edu/ofa/development/flexibility/part-time-guidelines/. As noted earlier in this report, when discussing part-time status, the committee is referring to work that is not contingent or marginalized in academia, but instead allows for a shift in intensity of work that is valued within the academy and can provide a path back to full-time work if desired.

[2] To learn more about the University of California system part-time program, see https://academicaffairs.ucdavis.edu/work-life.

[3] To learn more about the MIT Work Succeeding initiative, see https://hr.mit.edu/ws.

problem-solving leads to better outcomes, including validity of scientific findings. We have moved from the idealized image of the solo investigator generating brilliant new ideas to evidence that today's most effective science is in multiauthored reports (Maddi et al., 2023). Of great importance to the mission of this committee is recognizing that teams can absorb the variation in effort that may be required to juggle work and family.

The smallest team is a duo. Examples of two principal investigators sharing a laboratory include Nobel laureates Michael Brown and Joseph Goldstein and, more recently, Katalin Karikó and Drew Weissman. A joint laboratory structure can lead to a better distribution of workload and play to each person's strengths for a more fruitful creative process. It also provides trainees with different role models and perspective and, finally, helps even out the "ebb and flow of institutional knowledge" (Farese & Walther, 2021; Oldach, 2022).

Merging two existing laboratories or initiating a shared laboratory co-directed by two junior people avoids complications that could arise from a power differential. It is critical to get institutional acceptance of and willingness to support the scientific partnership model early in the process.

Job sharing is another alternative that allows meaningful engagement with decreased solo work responsibility. As a "creative approach to pursuing and achieving career goals for those with substantial obligations outside of their profession," two employees share the responsibilities of one full-time job (Sacks et al., 2015). For example, two clinicians each work 60 percent and cover each other on the 2 days the other is off. They share an office, so they are both in the office together 1 day a week and can reconnect and discuss patient issues, but they essentially have a private office on the other 2 of their 3 days in the office.

Administrative role sharing similarly provides flexibility. For example, two academic physician colleagues, a married couple who are colleagues, share the role of program director that allows one of them to decrease work hours and cover childcare (Sacks et al., 2015). In another example, two individuals with different areas of expertise each spend 60 percent time in sharing a vice chancellor's role. That leaves each 40 percent time for their own research and freedom to decrease total percent effort for a limited period when caregiving requires their attention. In addition to the flexibility benefit to each individual, the overlap uses the power and creativity of two people to brainstorm and is a bonus to the institution (Inge, 2018).

Open communication is essential in establishing and maintaining collaboration in job sharing, as well as any "nontraditional" work arrangement.

This collaboration must exist between individuals and also up the chain of command, as will be seen in the Predictability, Teaming, and Open Communication (PTO) model discussed later in this chapter. The initial design typically needs refining and can allow for modification over time as responsibilities outside or inside work change.

Grants to Support Caregiving Faculty

The flexibility necessitated by new caregiving responsibilities may mean cutting back from one's previous work-time commitment yet maintaining research productivity. Financial awards can help basic and clinical investigators maintain momentum as they integrate family caretaking responsibilities with job responsibilities. For example, funds may be used to offset the salaries of additional "hands" in the laboratory or clinic. Massachusetts General Hospital[4] and the Icahn School of Medicine at Mount Sinai[5] have instituted such programs to support their investigators who are adding caretaking to career commitments. These programs often are funded by individual philanthropy or institutional budgets and occasionally by foundations, such as the Alfred P. Sloan Foundation or Doris Duke Foundation (Jagsi et al., 2022; Jones et al., 2019, 2020; Szczygiel et al., 2021). (For information on the Doris Duke Fund to Retain Clinical Scientists see Box 6-4 in Chapter 6.)

Lehigh University used a Sloan Award for Faculty Career Flexibility to institute a program of faculty grants of $6,000 each "intended to help untenured tenure track faculty members sustain research productivity while caring for a newborn or adopted child, or other family member." Unlike the other faculty grants specifically targeted solely to supporting research, faculty members can use their grants in ways that they determine to be most useful, including covering costs for research travel or conferences, research assistance, technology, research materials, and for childcare and housekeeping.[6] While these grants no longer have Sloan funding, they have been institutionalized, showing one way that foundation support can ignite new workplace approaches for caregivers.

---

[4] To learn more about the Massachusetts General Hospital program, see https://ecor.mgh.harvard.edu/Default.aspx?node_id=226.

[5] To learn more about the Icahn School of Medicine program, see https://icahn.mssm.edu/about/gender-equity/programs.

[6] For more information about the Lehigh University grant program, see https://provost.lehigh.edu/sites/provost.lehigh.edu/files/Lehigh_Sloan_Research_Grant_Reimbursement_Guidelines.pdf.

Flexible Tenure Processes Through Areas of Excellence

SUNY Upstate Medical University takes a unique approach to tenure. Faculty are not limited to a single career track but instead are asked to identify an "area of excellence" after joining. They can choose among three options—research, clinical service, and education—and faculty are not restricted to a particular track based on the percentage effort spent in each. The aim of this approach is to allow flexibility to pursue opportunities in different areas as they arise, rather than being limited to only one domain.

Faculty members work with their department chair each year to establish their area of excellence through an "Annual Agreement of Academic Expectations." When a faculty member is up for promotion, they can select the area of excellence for the promotion committee to consider. In order to be promoted, the faculty member must demonstrate accomplishments in three domains in their area of excellence: leadership, innovation, and emerging regional reputation (American Council on Education et al., 2015).

Flexibility Supported by Time Banking

Given combined responsibilities of teaching, research, and clinical work, research has found that doctors on average work 10 hours more each week than other professionals (Shanafelt et al., 2012). This can result in stress that can lead to burnout, dissatisfaction with work, and ultimately, doctors choosing to leave. Though family-friendly policies may be officially available to help manage such stress, many physicians feel they cannot use them.

As noted in Chapter 5, one innovative solution is the time-banking system established at Stanford University School of Medicine. This chapter examines Stanford's program in more detail. The program was implemented under former Stanford University School of Medicine dean Phillip Pizzo based on the realization that many doctors were choosing to leave careers in academic medicine due to challenges balancing this with family needs. He established a task force composed of clinical and basic science faculty from all faculty tracks and ranks, ranging from instructor to full professor, to fully understand faculty needs and challenges regarding work-life fit, including a benchmarking survey across 10 leading academic medical centers, published research, and focus groups. However, this work did not yield creative or novel approaches, so the task force leaders sought help from the Stanford University Hasso Plattner Institute of Design, or d.school, and a human-centered design company in the San Francisco Bay Area, Jump Associates.

A key finding by the multidisciplinary Stanford-Jump team was that although Stanford had nearly every "family friendly" policy on the books, faculty were reluctant to use them to avoid "signaling low commitment" and the resulting adverse career consequences that have been reported in academic and other settings. The team concluded that cultural transformation was needed to gain acceptance of flexible work practices. They identified two major domains of conflict, work-life conflict and work-work conflict,[7] and created solutions for each conflict domain, structured around a framework they called Academic Biomedical Career Customization (ABCC) (Fassiotto et al., 2018).[8] Its guiding principles were that all participants have different needs, and that transparency is essential.

The Stanford-Jump team recommended the creation of a time-banking system to provide a means of recognizing faculty efforts to support the flexibility needed by their colleagues. This was incentivized through a system of credits accrued from efforts to support others' flexibility that allowed participants to "buy back" their time through the use of various services. The time faculty spent on service work that is often unappreciated, such as mentoring, committee service, and stepping in for colleagues on short notice, could be "banked" to receive credits that could then be used for services such as assistance with grant and manuscript writing, pre-made meals, housecleaning, care for children or older adults, and support for other household tasks such as with repairs or errands.

An evaluation of the program found that participants not only benefited in flexibility, but also saw workplace benefits. Specifically, results found that program participations received 1.3 times more grants during the period of analysis, resulting in over $1 million more in awarded funding per person (Fassiotto et al., 2018). These results suggest that innovative programs can help to reduce the extreme time pressures faced by academic medical faculty, even as institutional structures can remain restrictive. Stanford's ABCC program was part of a 2-year, $250,000 pilot funded largely by the Alfred P. Sloan Foundation.

---

[7] *Work-work conflict* was defined in the report as challenges caused by the competing demands academic medical faculty faced from different aspects of their jobs, including research, teaching, clinical obligations, service, and administrative tasks.

[8] The Academic Biomedical Career Customization planning tool created by Stanford University School of Medicine is available at https://sm.stanford.edu/app/abcc/.

Reentry Programs

Family caregivers may need to take time away from paid employment to focus on the needs of those that they care for, but reentry into the workforce can be a challenge, especially after an extended period away. Reentry programs provide on-ramps back into the paid labor force for those who needed extended breaks for any variety of reasons, including caregiving. Returnship and return-to-work programs generally provide training and mentorship to ease the transition back to work over the course of a few months and can help to encourage hiring managers to look beyond career gaps when considering potential hires (Vasel, 2021). Recent evidence suggests that men, however, may face penalties for using reentry programs due to gendered stigma against men leaving the workforce for childcare. Therefore, attention may be needed to ensure these programs are equally beneficial to everyone (Melin, 2023).

One example is the Re-Ignite program at Johnson & Johnson, which is aimed at experienced professionals ready to return to work after a career break of 2 years or longer. The program was started based on the realization that allowing for an on-ramp back into the workforce provided access to people who "come back to the workforce stronger and more prepared to take on the world's most critical health challenges" (Johnson & Johnson, 2018).

In 2015, the STEM Reentry Task Force microsite, a career reentry initiative, was started by the Society of Women Engineers and iRelaunch. The initiative aims to increase the representation of women in technical positions through supporting reentry through returnships and return-to-work programs. Their website features many companies that have job opportunities for reentry (STEM Reentry Task Force, 2015).

Directly relevant to restarting a career in academia are the National Institutes of Health (NIH) reentry supplement grants, which were established by the Office of Research on Women's Health in 1992. These grants were created to provide mentorship and retraining either full- or part-time for individuals looking to reenter an active research career following an interruption for caregiving or other qualifying events. Grant money can be used to update and extend research skills and to help reestablish the work they had been doing prior to a career interruption. Today, more than 20 of the institutes within NIH offer these grants (National Institutes of Health, 2023).

Predictability, Teaming, and Open Communication

Many professionals work 80-hour weeks both in and outside of STEMM. The result is often employee burnout, prompting exit from the organization, to search for something less stressful. An internal study conducted by a leading consulting firm, Boston Consulting Group (BCG), found that consultants could live with long hours but not with the unpredictability of those hours. BCG (working with Professor Leslie Perlow at Harvard Business School) developed a program entitled PTO—Predictability, Teaming and Open Communication—that is designed to rethink work processes and make the work more meaningful and manageable and thus reduce burnout (Kupp, 2021). This was a bottom-up initiative that began from asking and listening to their employees.

The three components of PTO are as follows:

1. *Predictability*: consistent, protected offline time over the course of each week.
2. *Teaming*: team collaboration and clear team norms to ensure everyone can take time off.
3. *Open communication*: regular conversations facilitated by an outside coach to raise and address key issues as soon as possible.

At BCG, PTO increased work-life satisfaction and led to a 74 percent increase in intention to stay with the firm for the long term. Moreover, consultants using PTO felt more able to manage a high-intensity, high-growth career. One ingredient to the success of PTO is that a "confidential coach" is assigned to each team to coach them through the process, serve as a mediator when disagreements arise, and assist in setting "team norms," such as not working on weekends. These coaches are already BCG employees and are selected based on their high emotional intelligence.

The principles of PTO can be appropriate for academia. By defining certain boundaries around work hours, such as no meetings after 5:00 p.m. or on weekends, and enabling an open collaborative effort, employees can be helped to achieve a more sustainable work-life balance, and junior colleagues can be empowered to advocate for themselves. Also, by working in teams, people are encouraged to communicate their individual scheduling expectations and check in regularly with team members. To the extent possible, having coaches available would facilitate the process.

Four-Day Workweek

The 5-day workweek was popularized by Henry Ford in 1926, with the goal of increasing production, and it has defined the workplace ever since. The assumption that the 5-day workweek is the most productive and efficient way to work has, however, been recently challenged by the results of a series of 4-day workweek trials. These trials were conducted over the last 18 months in multiple industries and sectors in the United States, Canada, the United Kingdom, and Ireland. Employees who worked 12 months at 4 days/full pay reported less burnout, improved mental and physical health, and better work-life balance. Employers who adopted these 4-day workweeks reported decreased turnover, sustained productivity, and because communication stayed constant, clients often did not even notice that the firms worked a 4-day week. To achieve these results, the 4-day workweek was the standard schedule within an organization, and old ways of working had to be rethought. For example, in many of the participating organizations, the number of weekly meetings was reduced and uninterrupted time to focus was increased. "The longer people worked in new, more efficient ways, the shorter their workweek became" (Fuhrmans, 2023; Thomas, 2021).

Bring Your Baby to Work

Another innovation that came from *asking and listening* is the Babies at Work program at Badger, a New Hampshire business that produces organic skin care products. In 2008, an employee who wanted to bring her new baby to work prompted Badger to work with the Parenting in the Workplace Institute to study other programs regarding babies in the workplace. Badger's resulting program, still in place, allows parents to bring their infants up to 6 months old to work.[9]

Prior to the baby's arrival, Badger's Human Resources department works with the parent-to-be to develop a Memo of Understanding that lays out business expectations, specifying the number of daily paid work hours and identifying the backup person to care for the baby when work necessitates that the parent step away. During the period that the baby comes to work, parents typically continue to work 8-hour days, but are paid for 6 or

---

[9] To learn more about Badger's program, see https://www.badgerbalm.com/pages/babies-at-work.

fewer hours, since everyone realizes no one can work 8 hours straight and care for an infant. The company reports that they have no recruitment costs, that turnover is very low, and that employee engagement is, in their words, "through the roof" (W. S. Badger Company, 2016).

SUMMARY OF FINDINGS FROM CHAPTER 7

Though the programs presented in this chapter may not have the rigorous research backing of previously discussed best practices, they represent unique and innovative approaches to the challenges faced in academic STEMM and other similar workplaces where cultural norms present barriers to true flexibility. Additionally, for many of these programs, anecdotal outcomes as well as some early examinations have been positive. If this report is to advance thinking and support for caregivers, it is imperative to think beyond existing approaches. COVID-19 has shown that American workplaces can rethink their time and timing of work, as well as their work locations and processes, to create more flexibility for employees and enhanced outcomes for employers. There is growing recognition that the structure of when, where, and how we work is mismatched to the needs of an increasingly diverse workforce (Christensen & Schneider, 2010).

1. Current innovations allow us to consider what the future may look like to better meet the needs of family caregivers.
2. Innovative solutions exist across a variety of sectors, including academia, nonprofit organizations, and for-profit businesses that seek to challenge existing norms and practices that hinder family caregivers.
3. Innovative solutions range from simple shifts and accommodations such as fostering shared laboratories and positions allowing individuals to bring their babies to work to a more fundamental rethinking of work in the form of a 4-day workweek.
4. Innovative solutions should aim to build trust, transparency, asking and listening, collaboration, and/or accountability to help promote a more caregiver-friendly environment.

8

Recommendations and Conclusions

Improving support for family caregivers in academic science, technology, engineering, mathematics, and medicine (STEMM) is vital for improving equity, strengthening innovation and creativity in science, preventing workforce shortages and critical skill gaps, and creating a more flexible and inclusive environment for all scientists. These goals require action at multiple levels and from various groups: colleges and universities, federal agencies and other funders, and federal and state governments. In this chapter, the committee outlines its recommendations for each.

RECOMMENDATIONS FOR COLLEGES AND UNIVERSITIES

Colleges and universities that aspire to support caregivers among their workforce and student body have many opportunities to enact, revise, publicize, improve implementation of, and extend policies and programs. Without these intentional actions, universities risk turnover, failure to recruit, and failure to retain top talent among those with caregiving responsibilities. We present recommendations in categories that represent distinct stages of action, from legal compliance to best practices, and finally, to innovative actions. We encourage colleges and universities to review their current practices, identify opportunities for implementation and growth, and publicly commit to improvement.

The overarching goal of these recommendations is to help universities create an environment that allows for continued and sustainable productivity in a way that is more inclusive of family caregivers. Such an environment shows a continued commitment to the long-term health and well-being of the academic STEMM workforce and challenges ideals of overwork as well as barriers to needed leave and flexibility. This overarching goal is reflected throughout these recommendations, which provide individual, concrete steps that can be taken and together can serve to shift broader cultural norms in more inclusive ways.

Legal Compliance

First, and most importantly, colleges and universities need to adopt effective measures to ensure that they protect caregivers' rights under current federal, state, and local laws. Under existing laws, students, staff, and faculty are typically entitled to leave, accommodations and work alterations, nursing/pumping facilities and accommodations, and nondiscrimination. However, the legal framework is fragmented and complicated, which contributes to a lack of awareness and compliance, as detailed in Chapter 6.

RECOMMENDATION 1: To ensure accountability and compliance, college and university leadership need to appoint a senior leader, ombuds, or team who is responsible for protecting, publicizing, and monitoring compliance with the legal mandates under Title IX, Title VII, the Family Medical and Leave Act (FMLA), the Pregnant Workers Fairness Act, and any state- and local-level policies that protect caregiving faculty, postdocs and other trainees, students, and staff by adopting the following practices:

   a. Ensure that requests for leave and accommodation and complaints of discrimination are handled by a specially trained, institution-wide administrator or administrators working together, and not assigned to departmental personnel.
   b. Provide leave for birthing parents to allow time for the birthing parent's physical recovery, inclusive of students and postdocs and other trainees. In some circumstances (described in Chapter 6), paid leave is required.
   c. Provide school or work accommodations for students, postdocs and other trainees, faculty, and staff who have needs related to pregnancy, childbirth, adoption, fostering, older adult care, or care related to a family member's physical or mental health.

d. Provide time and readily accessible space for pumping for breastfeeding/lactating individuals and ensuring that time for pumping is provided without penalty for students, postdocs and other trainees, faculty, and staff.
   e. Train Title IX officers, faculty, and department chairs so that they fully understand and support the legal rights of caregivers.
   f. Train faculty and administrators that it is illegal to make anyone "pay back" a leave and illegal to require anyone who is on paid or unpaid leave to work.
   g. Outline a clear process to file complaints that applies to individuals who believe their rights have been violated and ensure that complaints are resolved in a timely manner.

Best Practices

As outlined in Chapters 4 and 6, current policies supporting caregivers encompass leave, accommodations and adjustments, and direct care support. Colleges and universities need to adopt best practices in each area to ensure that family caregivers can fully participate in their scientific roles. To be most effective, caregiving leave policies need to extend well beyond what is required by FMLA and provide paid leave to all employees and ensure that students are not penalized for taking leave. Accommodations and adjustments should be institutionalized as a strategy to improve the support and flexibility needed by students and employees. Finally, direct care support should be centralized to make it easier to access and understand the available resources.

In the absence of these best practices, legal compliance can be implemented in a way that inequalities remain or are even exacerbated rather than mitigated. For example, as discussed in Chapter 4, universal and opt-out caregiving policies more effectively increase representation especially of women of color, while opt-in policies that are not universal may meet legal requirements but do not promote equity in the same way. These best practices are important not only to ensure effective support for family caregivers but also to bolster the positive effect of legal compliance. The committee also recommends continued data collection and analysis to ensure policy efficacy and address any unintended consequences.

RECOMMENDATION 2: *Caregiving Leave.* **Colleges and universities should comply with FMLA's requirement for 12 weeks of unpaid leave**

per year and provide paid family and medical leave to faculty, staff, postdocs and other trainees, and graduate students receiving pay, even if this leave is not mandated by state or federal law. Additionally, colleges and universities should provide leave for caregiving students, which allows them to maintain their student status so that they can continue to receive any aid or health insurance to which they are entitled. In developing their leave policies, colleges and universities need to consider the following:

  a. To build on the best practice of 12 weeks of paid leave for faculty for childbearing and child bonding, colleges and universities should consider similar provisions for other members of the academic scientific workforce, including staff, postdocs and other trainees, and graduate students receiving pay.[1]
  b. Develop creative funding solutions to extend the definition of caregiving leave to encompass all contexts of caregiving (care for adult children, older adults, extended family and kin, etc.).
  c. Provide guidance to students taking academic leave on whether taking leave will require their training period to be extended.

RECOMMENDATION 3: *Accommodations and adjustments.* **Colleges and universities should institutionalize opportunities for individually customized work and educational flexibility across a variety of needs, including location, time, workload, and intensity. In doing so, colleges and universities need to adopt the following practices:**

  a. Ensure equitable access to accommodations and alternative work and educational arrangements across all groups of employees and trainees, including postdocs, who are vulnerable to falling through administrative cracks at some academic institutions where they are categorized as neither employees nor students.
  b. Consider reduced load or part-time appointments (pre- and post-tenure) for faculty who have caregiving responsibilities that allow for transitions back to full-time work and facilitate increasing or decreasing professional effort over the course of a career.

---

[1] The committee acknowledges that some institutions may face strong financial constraints that make this fiscally infeasible. Colleges and universities operating under such constraints should aim to provide the greatest support possible and seek out alternative funding methods to increase support in the future.

RECOMMENDATIONS AND CONCLUSIONS 139

c. Implement flexible education policies for students, such as priority registration and part-time enrollment options for students who have caregiving needs, as well as options for excused absences or remote attendance due to caregiving responsibilities.
d. Ensure that engaging in flexibility through adapting location, time, or intensity in work or education to address caregiving demands is not used in evaluations of faculty, staff, postdocs and other trainees, or students to deny promotions, educational advancement, or access to resources.
e. Require that policies which provide adjustments for caregiving needs, such as stop-the-clock policies, are only used for caregiving and not as a form of sabbatical.
f. Make caregiver-friendly policies opt out, not opt in, so they are automatic and apply to all contexts of caregiving, which has been shown to produce greater benefits particularly for women of color compared with opt-in policies.

RECOMMENDATION 4: *Direct care support.* **Centralized resources to support basic caregiving needs for staff, faculty, postdocs and other trainees, and students need to be easily available and searchable. The following considerations should guide the creation and dissemination of these resources:**

a. Ensure resources are written down, well publicized, and readily accessible both online and in a central human resources (HR) office where caregivers can ask questions, in confidence if requested, about their specific needs and situations.
b. Identify resources that are already offered across departments as well as programs that are relevant to family caregivers and ensure there are adequate referral processes and networks to connect caregiving students, postdocs and other trainees, faculty, and staff to these resources.
c. Provide training and easily accessible materials through centralized HR offices for department chairs and faculty advisors who may need to share information about policies to caregivers they manage/advise to ensure the information shared is accurate and accessible by all.
d. Engage department chairs and supervisors to disseminate information on available policies to support caregivers and initiate an

annual conversation with each relevant employee to consider their needs for flexibility and discuss how their needs could be met, rather than placing the onus on employees.
e. Consider multiple forms of direct care support, including on-site care as well as care subsidies or reimbursements, to support those who may prefer to provide care themselves or seek trusted others rather than accessing on-site care due to cultural preferences or past experiences of discrimination.
f. Consider all caregiving contexts when developing direct care support, with particular attention to those that are often overlooked, including older adult care, care for extended family, and care for neurodivergent or disabled children.

RECOMMENDATION 5: *Data Collection and Analysis.* **To ensure that colleges and universities understand the needs of the caregiving populations within their ranks, understand the impact of their policies, existing and new, and address potential unintended consequences, colleges and universities should collect and analyze data on family caregivers. This should be accomplished through the following actions:**

a. Require relevant offices (e.g., offices of institutional research, human resources, offices of diversity equity and inclusion, provosts' offices, offices of student success, offices of financial aid) to expand existing climate surveys to include a standardized instrument on caregiving to collect data on the number of faculty, students, postdocs and other trainees, and staff with caregiving responsibilities, attitudes toward caregivers and caregiving, and impacts of current caregiving policies on those with and without caregiving responsibilities.
b. Require that relevant offices at colleges and universities ensure rigorous data collection and assessment of potential positive and negative effects of caregiving policies and accommodations. These offices should prioritize multiple methods of data collection, including qualitative interviews to understand the benefits and consequences of current policies.
c. Examine hiring and promotion specifically to ensure accountability and transparency in these processes that are central to ensuring the fair treatment and advancement of caregiver employees.

Innovative Practices

While best practices can be effectively implemented to support caregivers, there is a persistent need for innovation, particularly to address the pressing need for cultural change to better support effective policies as well as to develop new and cutting-edge practices.

RECOMMENDATION 6: Colleges and universities should pilot and evaluate innovative policies and practices intended to increase support for caregivers and influence lasting cultural change. Less research-intensive colleges and universities should partner with research-intensive institutions and participate in projects and efforts to test new policy ideas. For example, colleges and universities should consider the following:

a. Initiate easily implemented actions that normalize family caregiving, such as
   i. Fostering the creation of affinity groups and peer support groups for caregivers that could serve as a forum for making recommendations to institutional leaders.
   ii. Providing opportunities for leaders, including college and university presidents, provosts, and department chairs, to normalize conversations about caregiving as a natural part of life navigated by all.
   iii. Showcasing caregivers, including university leadership, engaged in various forms of caregiving in university communications and resources, with a focus on forms of caregiving that are often overlooked, such adult dependent care, older adult care, and care for extended family and loved ones.
   iv. Enhancing visibility and access to information about caregiving policies and infrastructure (e.g., day care centers, lactation pods) on websites and within campus offices.
b. Engage new and creative solutions from both within and outside academia to promote a culture of greater flexibility, such as
   i. Team-based science and teaching, flexible tenure processes, time-banking programs, temporary changes in professional effort, reentry programs, and modified work schedules.

RECOMMENDATIONS FOR FEDERAL AND PRIVATE FUNDERS

Along with universities, federal agencies and other funders play a major role in supporting the research performed at universities across the country as well as the researchers who conduct this work. To ensure that researchers who have caregiving responsibilities can effectively use grant funding, federal and private funders should focus on three key goals: (1) allow and support flexibility, particularly in the timing of grant eligibility and grant deadlines; (2) assist in leave and reentry; and (3) fund innovative research on family caregiving and use this research to develop and disseminate caregiving policy guidance to the institutions they fund. Many of these points have also been discussed by practitioners and experts in family caregiver support (Torres et al., 2023a, 2023b).

RECOMMENDATION 7: Federal and private funders should allow and support flexibility in the timing of grant eligibility as well as grant application and delivery deadlines for those with caregiving responsibilities and provide support for coverage while a grantee is on caregiving leave.[2] **Funders can implement this through the following actions:**

a. Decrease and streamline the paperwork and approval processes for grant applications.
b. Allow no-cost grant extensions based on caregiving needs.
c. Provide flexibility in eligibility timelines when an investigator has taken a caregiving leave, such as eligibility deadlines for early-career scholars.
d. Consider caregiving leave and acute caregiving demands as valid reasons for acceptance of a late application along the same timelines as other late applications.
e. Introduce and allow grant supplements or the redistribution of funding within a grant budget to support coverage for someone to continue scholarly work while the grantee is on caregiving leave

---

[2] When discussing coverage for grantees on caregiving leave, the committee is referring to coverage in research settings receiving outside funding from federal agencies or private funders. The committee acknowledges that staffing coverage presents a distinct and unique challenge in clinical care settings and encourages institutions to carefully assess their staffing needs and design robust systems to provide coverage readily when individuals with clinical service responsibilities require them.

as well as to provide support for caregiving-related expenses for conference and other research travel.

RECOMMENDATION 8: Federal and private funders should facilitate the leave and reentry processes for those who take a caregiving leave. In doing so, federal and private funders should take the following actions:

a. Provide research supplements to promote reentry following a period of caregiving leave.
b. Make supplements available to all types of caregivers, not solely parents, and cover costs associated with restarting a laboratory or research program as well as professional retraining.

RECOMMENDATION 9: Federal and private funders should fund innovative research on family caregiving in academic STEMM by providing competitive grants to institutions to support pilot projects and develop policy innovations. Funders should collaboratively develop and offer caregiver policy guidance to the institutions they fund based on the findings of this research as well as existing evidence. In doing so, funders should take the following actions:

a. Ensure the efficacy and impact of these innovative programs is scientifically evaluated.
b. Ensure these grants provide resources for universities to organize virtual and in-person conferences to share best practices in supporting STEMM caregiving jointly with career support to disseminate knowledge on best practices.
c. Create platforms to recognize universities that have implemented best practices to support STEMM caregivers through awards and invite these institutions to speak at conferences sponsored by funding agencies to highlight their excellence and provide recognition.

RECOMMENDATIONS TO CONGRESS AND THE FEDERAL GOVERNMENT

The federal government plays a critical role in establishing the expectations for supporting family caregivers across the country and in academic STEMM. This role is crucial to advance growth and innovation in the United States and advancing workforce inclusion. The federal government also has the opportunity to enhance the global competitiveness of the U.S. labor market supports by joining the ranks of all other Organisation for Economic Co-operation and Development nations by providing national, paid caregiving leave. The federal government should focus on two primary goals: (1) provide 12 weeks of paid, comprehensive caregiving leave and (2) provide incentives to support caregiving in STEMM legislation.

RECOMMENDATION 10: Congress should enact legislation to mandate a minimum of 12 weeks of paid, comprehensive caregiving leave. This leave should cover various contexts of caregiving, including childcare, older adult care, spousal care, dependent adult care, extended family care, end-of-life care, and bereavement care.

RECOMMENDATION 11: Following the model of the recent CHIPS and Science Act, which required the provision of on-site childcare for those seeking access to funds supporting semiconductor development, the agency or department tasked with implementation of future STEMM-funding legislation should include support for childcare in the application requirements.

References

AARP. (2021). *Caregiving Out-of-Pocket Costs Study*. https://www.aarp.org/content/dam/aarp/research/surveys_statistics/ltc/2021/family-caregivers-cost-survey-2021.doi.10.26419-2Fres.00473.001.pdf

AARP & National Alliance for Caregiving. (2020). *Caregiving in the U.S.* https://www.aarp.org/content/dam/aarp/ppi/2020/05/full-report-caregiving-in-the-united-states.doi.10.26419-2Fppi.00103.001.pdf

Abelson, R., & Rau, J. (2023). Facing financial ruin as costs soar for elder care. *The New York Times*. https://www.nytimes.com/2023/11/14/health/long-term-care-facilities-costs.html

ABMS. (2021). *American Board of Medical Specialties Policy on Parental, Caregiver and Medical Leave During Training*. American Board of Medical Specialties. https://www.abms.org/wp-content/uploads/2020/11/parental-caregiver-and-medical-leave-during-training-policy.pdf

Aburto, J. M., Tilstra, A. M., Floridi, G., & Dowd, J. B. (2022). Significant impacts of the COVID-19 pandemic on race/ethnic differences in US mortality. *Proceedings of the National Academy of Sciences*, *119*(35). https://doi.org/10.1073/pnas.2205813119

Acker, J. (1990). Hierarchies, jobs, bodies: A theory of gendered organizations. *Gender & Society*, *4*(2), 139-158. https://doi.org/10.1177/089124390004002002

Acker, J. (2006). Inequality regimes: Gender, class, and race in organizations. *Gender & Society*, *20*(4), 441-464. https://doi.org/10.1177/0891243206289499

Administration for Community Living. (2022). *2022 National Strategy to Support Family Caregivers*. https://acl.gov/CaregiverStrategy

American Association of Colleges of Nursing. (2023). *New Data Show Enrollment Declines in Schools of Nursing, Raising Concerns About the Nation's Nursing Workforce*. https://www.aacnnursing.org/news-data/all-news/new-data-show-enrollment-declines-in-schools-of-nursing-raising-concerns-about-the-nations-nursing-workforce

American Council on Education, Alfred P. Sloan Foundation, & NENFA. (2015). *Career Flexibility for Biomedical Faculty of Today and Tomorrow: Executive Summary*. Career Flexibility for Biomedical Faculty of Today and Tomorrow, Boston, MA.

American Society for Cell Biology. (2022). *Cell Bio Travel Grants*. American Society for Cell Biology and European Molecular Biology Organization. https://www.ascb.org/cellbio2022/register/travelgrants/

Andersen, J. P., Nielsen, M. W., Simone, N. L., Lewiss, R. E., & Jagsi, R. (2020). COVID-19 medical papers have fewer women first authors than expected. *eLife, 9*. https://doi.org/10.7554/eLife.58807

Anderson, D. J., Binder, M., & Krause, K. (2002). The motherhood wage penalty: Which mothers pay it and why? *American Economic Review, 92*(2), 354-358. https://doi.org/10.1257/000282802320191606

Anderson, M., & Goldman, R. H. (2020). Occupational reproductive hazards for female surgeons in the operating room: A review. *JAMA Surgery, 155*(3), 243-249. https://jamanetwork.com/journals/jamasurgery/article-abstract/2757728

Antecol, H., Bedard, K., & Stearns, J. (2018). Equal but inequitable: Who benefits from gender-neutral tenure clock stopping policies? *American Economic Review, 108*(9), 2420-2441. https://doi.org/10.1257/aer.20160613

Anthony, D. J. (2011). Tradition, conflict, and progress: A closer look at childbirth and parental leave policy on university campuses. *Georgetown Journal of Gender and the Law, 12*, 91.

AP-NORC Center for Public Affairs Research. (2014). *Long-Term Care in America: Expectations and Reality*. https://www.longtermcarepoll.org/wp-content/uploads/2017/11/AP-NORC-Long-term-Care-2014_Trend_Report.pdf

Archer, J., Reiboldt, W., Claver, M., & Fay, J. (2021). Caregiving in quarantine: Evaluating the impact of the Covid-19 pandemic on adult child informal caregivers of a parent. *Gerontology and Geriatric Medicine, 7*, 2333721421990150. https://doi.org/10.1177/2333721421990150

Armenia, A., & Gerstel, N. (2006). Family leaves, the FMLA and gender neutrality: The intersection of race and gender. *Social Science Research, 35*(4), 871-891. https://doi.org/https://doi.org/10.1016/j.ssresearch.2004.12.002

Arora, V. M., Wray, C. M., O'Glasser, A. Y., Shapiro, M., & Jain, S. (2020). Using the curriculum vitae to promote gender equity during the COVID-19 pandemic. *Proceedings of the National Academy of Sciences, 117*(39), 24032-24032. https://doi.org/10.1073/pnas.2012969117

Aughinbaugh, A., & Woods, R. A. (2021). *Patterns of Caregiving and Work: Evidence from Two Surveys*. https://www.bls.gov/opub/mlr/2021/article/patterns-of-caregiving-and-work-evidence-from-two-surveys.htm

Azmat, G., & Ferrer, R. (2017). Gender gaps in performance: Evidence from young lawyers. *Journal of Political Economy, 125*(5), 1306-1355. https://doi.org/10.1086/693686

Bainbridge, H. T. J., & Broady, T. R. (2017). Caregiving responsibilities for a child, spouse or parent: The impact of care recipient independence on employee well-being. *Journal of Vocational Behavior, 101*, 57-66. https://doi.org/https://doi.org/10.1016/j.jvb.2017.04.006

Baldiga, M., Joshi, P., Hardy, E., & Acevedo-Garcia, D. (2018). *Data-for-Equity Research Brief: Child Care Affordability for Working Parents*. https://www.diversitydatakids.org/research-library/research-brief/child-care-affordability-working-parents

Beckerle, M. C., Reed, K. L., Scott, R. P., Shafer, M.-A., Towner, D., Valantine, H. A., & Zahniser, N. R. (2011). Medical faculty development: A modern-day odyssey. *Science Translational Medicine, 3*(104), 104cm131. https://doi.org/doi:10.1126/scitranslmed.3002763

Benard, S., Paik, I., & Correll, S. J. (2004). Cognitive bias and the motherhood penalty. *UC Law Journal*. https://repository.uclawsf.edu/hastings_law_journal/vol59/iss6/3/

Bender, S., Brown, K. S., Hensley Kasitz, D. L., & Vega, O. (2022). Academic women and their children: Parenting during COVID-19 and the impact on scholarly productivity. *Family Relations, 71*(1), 46-67. https://doi.org/10.1111/fare.12632

Berberet, J., Bland, C. J., Brown, B. E., & Risbey, K. R. (2005). *Late Career Faculty Perceptions: Implications for Retirement Planning and Policymaking*. https://www.tiaa.org/public/institute/publication/2005/late-career-faculty-perceptions-implications

Berg, S. (2018). *Working Overtime? At Stanford, Physicians Bank the Time for Later*. https://www.ama-assn.org/practice-management/physician-health/working-overtime-stanford-physicians-bank-time-later

Bertrand, M., Goldin, C., & Katz, L. F. (2010). Dynamics of the gender gap for young professionals in the financial and corporate sectors. *American Economic Journal: Applied Economics, 2*(3), 228-255. https://doi.org/10.1257/app.2.3.228

A Better Balance. (2023). Comparative Chart of Paid Family and Medical Leave Laws in the United States.

Bianchi, S. M., Robinson, J. P., & Milke, M. A. (2006). *The Changing Rhythms of American Family Life*. Russell Sage Foundation.

Biblarz, T. J., & Savci, E. (2010). Lesbian, gay, bisexual, and transgender families. *Journal of Marriage and Family, 72*(3), 480-497. https://doi.org/10.1111/j.1741-3737.2010.00714.x

Bilmes, L. (2021). The Long-Term Costs of United States Care for Veterans of the Afghanistan and Iraq Wars. *HKS Faculty Research Working Paper Series*. https://watson.brown.edu/costsofwar/files/cow/imce/papers/2021/Costs%20of%20War_Bilmes_Long-Term%20Costs%20of%20Care%20for%20Vets_Aug%202021.pdf

Bird, S. R. (2011). Unsettling universities' incongruous, gendered bureaucratic structures: A case-study approach. *Gender, Work & Organization, 18*(2), 202-230. https://doi.org/10.1111/j.1468-0432.2009.00510.x

Blair-Loy, M. (2001). Cultural constructions of family schemas: The case of women finance executives. *Gender & Society, 15*(5), 687-709.

Blair-Loy, M., & Cech, E. (2022). *Misconceiving Merit: Paradoxes of Excellence and Devotion in Academic Science and Engineering*. The University of Chicago Press.

Blair-Loy, M., & Cech, E. A. (2017). Demands and devotion: Cultural meanings of work and overload among women researchers and professionals in science and technology industries. *Sociological Forum, 32*(1), 5-27. http://www.jstor.org/stable/26626058

Blair-Loy, M., Reynders, S., & Cech, E. A. (2023). Productivity metrics and hiring rubrics are warped by cultural schemas of merit. *Trends in Microbiology, 31*(6), 556-558. https://doi.org/10.1016/j.tim.2023.03.004

Body, D. (2020). *The True Cost of Caregiving*. https://www.aspeninstitute.org/wp-content/uploads/2020/05/The-True-Cost-of-Caregiving.pdf

Boelmann, B., Raute, A., & Schonberg, U. (2021). Wind of change? Cultural determinants of maternal labor supply. *SSRN Electronic Journal*. https://doi.org/10.2139/ssrn.3855973

Bose, M., Tokarewich, L., Bratches, R. W. R., & Barr, P. J. (2021). *Caregiving in a Diverse America: Beginning to Understand the Systemic Challenges Facing Family Caregivers*. https://www.caregiving.org/wp-content/uploads/2021/11/NAC_AmgenDiverseCaregiversReport_FinalDigital-111821.pdf

Boss, P., Bryant, C., & Mancini, J. (2017). *Family Stress Management: A Contextual Approach*. https://doi.org/10.4135/9781506352206

Boyd, K., Winslow, V., Borson, S., Lindau, S. T., & Makelarski, J. A. (2022). Caregiving in a pandemic: Health-related socioeconomic vulnerabilities among women caregivers early in the COVID-19 pandemic. *Annals of Family Medicine, 20*(5), 406-413. https://doi.org/10.1370/afm.2845

Brown, V., & Nichols, T. R. (2013). Pregnant and parenting students on campus: Policy and program implications for a growing population. *Educational Policy, 27*(3), 499-530.

Buchanan, D., Fitzgerald, L., Ketley, D., Gollop, R., Jones, J. L., Lamont, S. S., Neath, A., & Whitby, E. (2005). No going back: A review of the literature on sustaining organizational change. *International Journal of Management Reviews, 7*(3), 189-205. https://doi.org/10.1111/j.1468-2370.2005.00111.x

Budig, M. J., & England, P. (2001). The wage penalty for motherhood. *American Sociological Review, 66*(2), 204-225. https://doi.org/10.2307/2657415

Burch, K. A., Sorensen, M. B., Hurt, C. E., Simmons, M. R., Eugene, T., McDaniel, A. K., & Paulson, A. (2023). Parental leave is just a wolf in sheep's clothing: A call for gender-aware policies in academia. *Industrial and Organizational Psychology, 16*(2), 277-282.

Butrica, B., & Karamcheva, N. (2015). *The Impact of Informal Caregiving on Older Adults' LaborSupply and Economic Resources*. https://www.urban.org/research/publication/impact-informal-caregiving-older-adults-laborsupply-and-economic-resources

Bye, E. M., Brisk, B. W., Reuter, S. D., Hansen, K. A., & Nettleman, M. D. (2017). Pregnancy and parenthood during medical school. *South Dakota Medicine, 70*(12).

Calarco, J. M., Meanwell, E., Anderson, E. M., & Knopf, A. S. (2021). By default: How mothers in different-sex dual-earner couples account for inequalities in pandemic parenting. *Socius: Sociological Research for a Dynamic World, 7*, 237802312110387. https://doi.org/10.1177/23780231211038783

Calisi, R. M., & Working Group of Mothers in Science. (2018). How to tackle the childcare–conference conundrum. *Proceedings of the National Academy of Sciences, 115*(12), 2845-2849. https://doi.org/10.1073/pnas.1803153115

Callahan, C., Boustani, M., Sachs, G., & Hendrie, H. (2009). Integrating care for older adults with cognitive impairment. *Current Alzheimer Research, 6*(4), 368-374. https://doi.org/10.2174/156720509788929228

Calvano, L. (2013). Tug of war: Caring for our elders while remaining productive at work. *Academy of Management Perspectives, 27*(3), 204-218. http://www.jstor.org/stable/43822022

REFERENCES

Calvert, C. T. (2016). *Caregivers in the Workplace: Family Responsibilities Discrimination Litigation Update 2016*. https://worklifelaw.org/publications/Caregivers-in-the-Workplace-FRD-update-2016.pdf

Caplan, Z., & Rabe, M. (2023). *The Older Population: 2020*. https://www.census.gov/library/publications/2023/decennial/c2020br-07.html

Cardel, M. I., Dhurandhar, E., Yarar-Fisher, C., Foster, M., Hidalgo, B., McClure, L. A., Pagoto, S., Brown, N., Pekmezi, D., Sharafeldin, N., Willig, A. L., & Angelini, C. (2020). Turning chutes into ladders for women faculty: A review and roadmap for equity in academia. *Journal of Women's Health, 29*(5), 721-733. https://doi.org/10.1089/jwh.2019.8027

Carr, P. L., Gunn, C., Raj, A., Kaplan, S., & Freund, K. M. (2017). Recruitment, promotion, and retention of women in academic medicine: How institutions are addressing gender disparities. *Women's Health Issues, 27*(3), 374-381. https://doi.org/10.1016/j.whi.2016.11.003

Carrigan, C., Quinn, K., & Riskin, E. A. (2011). The gendered division of labor among STEM faculty and the effects of critical mass. *Journal of Diversity in Higher Education, 4*(3), 131.

Carter, R. (2011). *Written Testimony of Former First Lady Rosalynn Carter Before the Senate Special Committee on Aging*. https://www.cartercenter.org/news/editorials_speeches/rosalynn-carter-committee-on-aging-testimony.html

Cech, E. A., & Blair-Loy, M. (2014). Consequences of flexibility stigma among academic scientists and engineers. *Work and Occupations, 41*(1), 86-110. https://doi.org/10.1177/0730888413515497

Cech, E. A., & Blair-Loy, M. (2019). The changing career trajectories of new parents in STEM. *Proceedings of the National Academy of Sciences, 116*(10), 4182-4187. https://www.pnas.org/doi/pdf/10.1073/pnas.1810862116

Centers for Disease Control and Prevention. (2019). *Caregiving for Family and Friends—A Public Health Issue*. U.S. Department of Health and Human Services. https://www.cdc.gov/aging/caregiving/caregiver-brief.html

Cha, Y., & Weeden, K. A. (2014). Overwork and the slow convergence in the gender gap in wages. *American Sociological Review, 79*(3), 457-484. https://doi.org/10.1177/0003122414528936

Child Care Aware of America. (2022). *Demanding Change: Repairing our Child Care System*. https://www.childcareaware.org/demanding-change-repairing-our-child-care-system/

Christensen, K. (2013). Launching the workplace flexibility movement: Work family research and a program of social change. *Community, Work & Family, 16*(3), 261-284. https://doi.org/10.1080/13668803.2013.820092

Christensen, K., & Schneider, B. (2010). *Workplace Flexibility: Realigning 20th-Century Jobs for a 21st-Century Workforce*. Cornell University Press. http://www.jstor.org/stable/10.7591/j.ctt7v9h8

Clancy, R. L., Fisher, G. G., Daigle, K. L., Henle, C. A., McCarthy, J., & Fruhauf, C. A. (2020). Eldercare and work among informal caregivers: A multidisciplinary review and recommendations for future research. *Journal of Business and Psychology, 35*(1), 9-27. https://doi.org/10.1007/s10869-018-9612-3

Clark, A. M., Lucero-Nguyen, Y., Cavuoti, B., Dale-Bley, S., Duran, Z., & Embry, C. (2021). *Toolkit for establishing lactation support on university and college campuses.* https://ksbreastfeeding.org/wp-content/uploads/2021/02/Lactation-Support-on-University-Campuses.pdf

Cohen, S. A., Nash, C. C., & Greaney, M. L. (2021). Informal caregiving during the COVID-19 pandemic in the US: Background, challenges, and opportunities. *American Journal of Health Promotion, 35*(7), 1032-1036. https://doi.org/10.1177/08901171211030142c

Cohen, S. A., Sabik, N. J., Cook, S. K., Azzoli, A. B., & Mendez-Luck, C. A. (2019). Differences within differences: Gender inequalities in caregiving intensity vary by race and ethnicity in informal caregivers. *Journal of Cross-Cultural Gerontology, 34*(3), 245-263. https://doi.org/10.1007/s10823-019-09381-9

Collins, P. H. (1991). Black women and motherhood. In V. Held (Ed.), *Justice and Care.* Westview Press.

Conley, C. S., Caldwell, M. S., Flynn, M., & Dupre, A. J. (2004). *Handbook of parenting: Theory and research for practice.* SAGE Publications Ltd. https://doi.org/10.4135/9781848608160

Contreras-Mendez, S., & Cruse, L. R. (2021). *Busy with Purpose: Lessons for Education and Policy Leaders from Returning Student Parents.* Institute for Women's Policy Research.

Coronel, B. (2020). The lived experience of community college student-parents. *Aleph, UCLA Undergraduate Research Journal for the Humanities and Social Sciences, 17*(1).

Correll, S., Benard, S., & Paik, I. (2007). Getting a job: Is there a motherhood penalty? *American Journal of Sociology, 112*(5), 1297-1338. https://doi.org/10.1086/511799

Cortés, P., & Pan, J. (2020). *Children and the Remaining Gender Gaps in the Labor Market.* https://dx.doi.org/10.3386/w27980

Cuddy, A. J. C., Fiske, S. T., & Glick, P. (2004). When professionals become mothers, warmth doesn't cut the ice. *Journal of Social Issues, 60*(4), 701-718. https://doi.org/10.1111/j.0022-4537.2004.00381.x

Culpepper, D., & Kilmer, S. (2022). Faculty-related COVID-19 policies and practices at top-ranked higher education institutions in the United States. *Advance Journal, 3*(2).

Cutler, E. (2023). True cost of childcare by state. TOOTRis. https://tootris.com/edu/blog/parents/cost-of-child-care-in-all-50-states-for-2022/

Cynkar, P., & Mendes, E. (2011). *More Than One in Six American Workers Also Act as Caregivers.* https://news.gallup.com/poll/148640/One-Six-American-Workers-Act-Caregivers.aspx

Czeisler, M., Rohan, E. A., Melillo, S., Matjasko, J. L., DePadilla, L., Patel, C. G., Weaver, M. D., Drane, A., Winnay, S. S., Capodilupo, E. R., Robbins, R., Wiley, J. F., Facer-Childs, E. R., Barger, L. K., Czeisler, C. A., Howard, M. E., & Rajaratnam, S. M. W. (2021). Mental health among parents of children aged <18 years and unpaid caregivers of adults during the COVID-19 pandemic—United States, December 2020 and February–March 2021. *Morbidity and Mortality Weekly Report, 70*(24), 879-887. https://doi.org/10.15585/mmwr.mm7024a3

Damaske, S., Ecklund, E. H., Lincoln, A. E., & White, V. J. (2014). Male scientists' competing devotions to work and family: Changing norms in a male-dominated profession. *Work and Occupations, 41*(4), 477-507. https://doi.org/10.1177/0730888414539171

Daskalska, L. L., O'Brien, B. S., Arzua, T., & Bakken, B. K. (2022). Family support policy for pharmacy, medical, and graduate students. *Journal of Science Policy & Governance.*

Davies, A. R., & Frink, B. D. (2014). The origins of the ideal worker: The separation of work and home in the United States from the Market Revolution to 1950. *Work and Occupations, 41*(1), 18-39. https://doi.org/10.1177/0730888413515893

Dembe, A. E., & Partridge, J. S. (2011). The benefits of employer-sponsored elder care programs: Case studies and policy recommendations. *Journal of Workplace Behavioral Health, 26*(3), 252-270. https://doi.org/10.1080/15555240.2011.589755

DeParle, J. (2021). When child care costs twice as much as the mortgage. *The New York Times*. https://www.nytimes.com/2021/10/09/us/politics/child-care-costs-wages-legislation.html

DesRoches, C. M., Zinner, D. E., Rao, S. R., Iezzoni, L. I., & Campbell, E. G. (2010). Activities, productivity, and compensation of men and women in the life sciences. *Academic Medicine, 85*(4), 631-639. https://doi.org/10.1097/ACM.0b013e3181d2b095

Dilworth-Anderson, P., Moon, H., & Aranda, M. P. (2020). Dementia caregiving research: Expanding and reframing the lens of diversity, inclusivity, and intersectionality. *Gerontologist, 60*(5), 797-805. https://doi.org/10.1093/geront/gnaa050

Dolamore, S., Henderson, A., & Carrizales, T. (2021). Structural obstacles for women in academia: Availability and costs of campus child care. *Journal of Public Management & Social Policy, 28*(1). https://digitalscholarship.tsu.edu/jpmsp/vol28/iss1/9

Donnelly, K., Twenge, J. M., Clark, M. A., Shaikh, S. K., Beiler-May, A., & Carter, N. T. (2016). Attitudes toward women's work and family roles in the United States, 1976–2013. *Psychology of Women Quarterly, 40*(1), 41-54. https://doi.org/10.1177/0361684315590774

Doren, C. (2019). Is two too many? Parity and mothers' labor force exit. *Journal of Marriage and Family, 81*(2), 327-344. https://doi.org/10.1111/jomf.12533

Douglas, H. M., Settles, I. H., Cech, E. A., Montgomery, G. M., Nadolsky, L. R., Hawkins, A. K., Ma, G., Davis, T. M., Elliott, K. C., & Cheruvelil, K. S. (2022). Disproportionate impacts of COVID-19 on marginalized and minoritized early-career academic scientists. *PLoS ONE, 17*(9), e0274278. https://doi.org/10.1371/journal.pone.0274278

Dow, D. M. (2015). Integrated motherhood: Beyond hegemonic ideologies of motherhood. *Journal of Marriage and Family, 78*(1), 180-196. https://doi.org/10.1111/jomf.12264

Drago, R., Colbeck, C. L., Stauffer, K. D., Pirretti, A., Burkum, K., Fazioli, J., Lazzaro, G., & Habasevich, T. (2006). The avoidance of bias against caregiving: The case of academic faculty. *American Behavioral Scientist, 49*(9), 1222-1247. https://doi.org/10.1177/0002764206286387

Drago, R., & Williams, J. (2000). A half-time tenure track proposal. *Change: The Magazine of Higher Learning, 32*(6), 46-51. https://doi.org/10.1080/00091380009601767

Durbin, S., & Tomlinson, J. (2010). Female part-time managers: Networks and career mobility. *Work, Employment and Society, 24*(4), 621-640. https://doi.org/10.1177/0950017010380631

Duxbury, L., & Dole, G. (2015). Squeezed in the middle: Balancing paid employment, childcare and eldercare. *Flourishing in life, work and careers: Individual wellbeing and career experiences* (pp. 141-166). Edward Elgar Publishing.

Duxbury, L., & Higgins, C. (2017). *Something's got to give: Balancing work, childcare and eldercare*. University of Toronto Press.

Eckerson, E., Talbourdet, L., Reichlin, L., Sykes, M., Noll, E., & Gault, B. (2016). *Child Care for Parents in College: A State-by-State Assessment.* https://iwpr.org/wp-content/uploads/2020/12/C445.pdf

Ecklund, E. H., Damaske, S., Lincoln, A. E., & White, V. J. (2017). Strategies men use to negotiate family and science. *Socius: Sociological Research for a Dynamic World, 3.* https://doi.org/10.1177/2378023116684516

Ecklund, E. H., & Lincoln, A. E. (2011). Scientists want more children. *PLoS ONE, 6*(8), e22590. https://doi.org/10.1371/journal.pone.0022590

Ecklund, E. H., & Lincoln, A. E. (2016). *Failing families, failing science: Work-family conflict in academic science.* New York University Press.

Ecklund, E. H., Lincoln, A. E., & Tansey, C. (2012). Gender segregation in elite academic science. *Gender & Society, 26*(5), 693-717. https://doi.org/10.1177/0891243212451904

England, P., & Srivastava, A. (2013). Educational differences in US parents' time spent in child care: The role of culture and cross-spouse influence. *Social Science Research, 42*(4), 971-988. https://doi.org/10.1016/j.ssresearch.2013.03.003

Englander, M. J., & Ghatan, C. (2021). Radiation and the pregnant IR: Myth versus fact. *CardioVascular and Interventional Radiology, 44,* 877-882. https://link.springer.com/content/pdf/10.1007/s00270-020-02704-1.pdf

Equal Employment Opportunity Commission. (1964). *Title VII of the Civil Rights Act of 1964.* U.S. Equal Employment Opportunity Commission.

Fahle, S., & McGarry, K. (2018). Caregiving and work: The relationship between labor market attachment and parental caregiving. *Innovation in Aging, 2*(Suppl 1), 580. https://doi.org/10.1093/geroni/igy023.2150

Farese, R. V., & Walther, T. C. (2021). The power of two: Lessons from a scientific partnership. *Journal of Clinical Investigation, 131*(2). https://doi.org/10.1172/jci145966

Fassiotto, M., Simard, C., Sandborg, C., Valantine, H., & Raymond, J. (2018). An integrated career coaching and time-banking system promoting flexibility, wellness, and success: A pilot program at Stanford University School of Medicine. *Academic Medicine, 93*(6), 881-887. https://doi.org/10.1097/acm.0000000000002121

Feeny, S., Iamsiraroj, S., & McGillivray, M. (2014). Remittances and economic growth: Larger impacts in smaller countries? *The Journal of Development Studies, 50*(8), 1055-1066. https://doi.org/10.1080/00220388.2014.895815

Ferguson, K., Chen, L., & Costello, T. (2021). *Growing Progress in Supporting Postdocs 2021: National Postdoctoral Association Institutional Policy Report.* https://www.sigmaxi.org/publications/postdoc-report

Fernandez, J., Reckrey, J., & Lindquist, L. A. (2016). Aging in place: Selecting and supporting caregivers of the older adult. *New Directions in Geriatric Medicine: Concepts, Trends, and Evidence-Based Practice* (pp. 115-126). https://www.cambridge.org/us/universitypress/subjects/psychology/developmental-psychology/formative-experiences-interaction-caregiving-culture-and-developmental-psychobiology?format=HB&isbn=9780521895033

Forry, N. D., & Hofferth, S. L. (2011). Maintaining work: The influence of child care subsidies on child care–related work disruptions. *Journal of Family Issues, 32*(3), 346-368. https://doi.org/10.1177/0192513x10384467

Fox, M., & Gaughan, M. (2021). Gender, family and caregiving leave, and advancement in academic science: Effects across the life course. *Sustainability, 13*(12), 6820. https://doi.org/10.3390/su13126820

REFERENCES   153

French, V. A., Werner, J. L., Feng, E. J. H., Latimer, R. A., Wolff, S. F., & Wieneke, C. L. (2022). Provision of onsite childcare in US academic health centers: What factors make a difference? *Women's Health Issues, 32*(1), 74-79. https://doi.org/https://doi.org/10.1016/j.whi.2021.08.005

Friedman, D. (2022). Feds in Gen X and Gen Z both want more work flexibility, but often for different reasons. *Federal News Network*. https://federalnewsnetwork.com/workforce/2022/11/feds-in-gen-x-and-gen-z-both-want-more-work-flexibility-but-often-for-different-reasons/

Fuentes-Afflick, E., García, P., Friedli, A. B., Johnson, B., & Binder, R. (2022). An innovative faculty travel award program to support child, elder, or dependent care. *JAMA Pediatrics, 176*(11), 1148. https://doi.org/10.1001/jamapediatrics.2022.2852

Fuhrmans, V. (2023). A four-day workweek experiment finds work does get done in less time. *The Wall Street Journal*. https://www.wsj.com/articles/the-four-day-workweek-gets-shorter-with-practice-companies-find-edc9eb2f

Future Forum. (2022). *Future Forum Pulse*. futureforum.com/pulse-survey

Gabriel, A. S., Allen, T. D., Devers, C. E., Eby, L. T., Gilson, L. L., Hebl, M., Kehoe, R. R., King, E. B., Ladge, J. J., Little, L. M., Ou, A. Y., Schleicher, D. J., Shockley, K. M., Klotz, A. C., & Rosen, C. C. (2023). A call to action: Taking the untenable out of women professors' pregnancy, postpartum, and caregiving demands. *Industrial and Organizational Psychology, 16*(2), 187-210. https://doi.org/10.1017/iop.2022.111

Galinsky, E., Bond, J. T., Sakai, K., Kim, S. S., & Giuntoli, N. (2008). *2008 National Study of Employers*. https://cdn.sanity.io/files/ow8usu72/production/b0bca7a-592140fa0000d7d8bbcd6ed13fbcfd759.pdf

Garfinkel, I., Rainwater, L., & Smeeding, T. (2010). *Wealth and Welfare States: Is America a Laggard or Leader?* Oxford University Press. https://EconPapers.repec.org/RePEc:oxp:obooks:9780199579310

Gatta, M. L., & Roos, P. A. (2004). Balancing without a net in academia: Integrating family and work lives. *Equal Opportunities International, 23*(3/4/5), 124-142. https://doi.org/10.1108/02610150410787765

Gault, B., Holtzman, T., & Cruse, L. R. (2020). *Understanding the Student Parent Experience: The Need for Improved Data Collection on Parent Status in Higher Education*. https://iwpr.org/wp-content/uploads/2020/10/Understanding-the-Student-Parent-Experience_Final.pdf

Gerstel, N. (2011). Rethinking families and community: The color, class, and centrality of extended kin ties. *Sociological Forum, 26*(1), 1-20. https://doi.org/10.1111/j.1573-7861.2010.01222.x

Gheyoh Ndzi, E. (2023). Paternal leave entitlement and workplace culture: A key challenge to paternal mental health. *International Journal of Environmental Research and Public Health, 20*(8), 5454. https://doi.org/10.3390/ijerph20085454

Global Coalition on Aging & Home Instead. (2021). *Building The Caregiving Workforce Our Aging World Needs*. https://globalcoalitiononaging.com/wp-content/uploads/2021/06/GCOA_HI_Building-the-Caregiving-Workforce-Our-Aging-World-Needs_REPORT-FINAL_July-2021.pdf

Goldin, C. (2023). *Why Women Won*. https://dx.doi.org/10.3386/w31762

Goldrick-Rab, S., Welton, C. R., & Coca, V. (2020). *Parenting while in college: Basic needs insecurity among students with children*. Hope Center for College, Community, and Justice. Temple University.

Gonsalves, L., Kalev, A. Dobbin, F., Kim, K. W., & Deutsch, G. (2022). How to Stop the Clock: Policy Universalism and the Effects of Tenure Clock Extensions on Faculty Diversity. Academy of Management Annual Meeting Proceedings, Seattle, WA.

Gonzalez, L., & Zoabi, H. (2021). Does paternity leave promote gender equality within households? *SSRN Electronic Journal*. https://doi.org/10.2139/ssrn.3971987

Goulden, M., Mason, M. A., & Frasch, K. (2011). Keeping women in the science pipeline. *The Annals of the American Academy of Political and Social Science*, *638*(1), 141-162. https://doi.org/10.1177/0002716211416925

Grigoryeva, A. (2017). Own gender, sibling's gender, parent's gender: The division of elderly parent care among adult children. *American Sociological Review*, *82*(1), 116-146. https://doi.org/10.1177/0003122416686521

Gromada, A., & Richardson, D. (2021). *Where Do rich Countries Stand on Childcare?* UNICEF. https://www.unicef-irc.org/publications/pdf/where-do-rich-countries-stand-on-childcare.pdf

Gulati, M., Korn, R. M., Wood, M. J., Sarma, A., Douglas, P. S., Singh, T., Merz, N. B., Lee, J., Mehran, R., & Andrews, O. A. (2022). Childbearing among women cardiologists: The interface of experience, impact, and the law. *Journal of the American College of Cardiology*, *79*(11), 1076-1087. https://www.sciencedirect.com/science/article/pii/S0735109722001693?via%3Dihub

Gunja, M. Z., Gumas, E. D., & Williams, R. D., II. (2022). *U.S. Health Care from a Global Perspective, 2022: Accelerating Spending, Worsening Outcomes*. Commonwealth Fund. https://www.commonwealthfund.org/publications/issue-briefs/2023/jan/us-health-care-global-perspective-2022

Guo, M., Kim, S., & Dong, X. (2019). Sense of filial obligation and caregiving burdens among Chinese immigrants in the United States. *Journal of the American Geriatrics Society*, *67*(S3). https://doi.org/10.1111/jgs.15735

Guo, M., LeRoux, M., & Lavery, L. (2023). *Postdoc Letter to NIH*. https://postdocparentsforchange.com

Halverson, C. (2003). From here to paternity: Why men are not taking paternity leave under the Family and Medical Leave Act. *Wisconsin Women's Law Journal*, *18*. https://repository.law.wisc.edu/s/uwlaw/item/24452

Hartmann, K. E., Sundermann, A. C., Helton, R., Bird, H., & Wood, A. (2018). The Scope of extraprofessional caregiving challenges among early career faculty: Findings from a university medical center. *Academic Medicine*, *93*(11), 1707-1712. https://doi.org/10.1097/acm.0000000000002229

Harvard Graduate Students Union. (2020). *Contract Summary*. https://harvardgradunion.org/contract-summary/

Hayes, T. O. N., & Kurtovic, S. (2020). The Ballooning Costs of Long-Term Care. *The American Action Forum*. https://www.americanactionforum.org/research/the-ballooning-costs-of-long-term-care/

He, D., & McHenry, P. (2015). Does formal employment reduce informal caregiving? *Health Economics*, *25*(7), 829-843. https://doi.org/10.1002/hec.3185

Henle, C. A., Fisher, G. G., McCarthy, J., Prince, M. A., Mattingly, V. P., & Clancy, R. L. (2020). Eldercare and childcare: How does caregiving responsibility affect job discrimination? *Journal of Business and Psychology*, *35*, 59-83.

Herr, J., Roy, R., & Klerman, J. A. (2020). *Gender Differences in Needing and Taking Leave*. Abt Associates Inc. https://www.dol.gov/sites/dolgov/files/OASP/evaluation/pdf/WHD_FMLAGenderShortPaper_January2021.pdf

Hill, C., & Rose, A. S. (2013). *Women in Community Colleges: Access to Success*. https://eric.ed.gov/?id=ED546790

Hoefman, R. J., Van Exel, J., & Brouwer, W. B. (2013). Measuring the impact of caregiving on informal carers: A construct validation study of the CarerQol instrument. *Health and Quality of Life Outcomes*, *11*(1), 173. https://doi.org/10.1186/1477-7525-11-173

Hollenshead, C. S., Sullivan, B., Smith, G. C., August, L., & Hamilton, S. (2005). Work/family policies in higher education: Survey data and case studies of policy implementation. *New Directions for Higher Education*, *2005*(130), 41-65. https://doi.org/https://doi.org/10.1002/he.178

Hosek, A. M., & Harrigan, M. M. (2023). Attributions and framing in working mothers' reports about division of family labor. *Journal of Family Communication*, *23*(1), 63-74. https://doi.org/10.1080/15267431.2023.2165080

Huerta, A. H., Salazar, M. E., Lopez, E. F., Torres, G., Badajos, L. M., & Lopez Matias, N. A. (2022). *Identifying Institutional Needs for Student Parents in Community Colleges: Recommendations for Successful Policy and Practice*. Pullias Center for Higher Education.

Ibrahim, H., Stadler, D. J., Archuleta, S., & Cofrancesco, J., Jr. (2017). Twelve tips to promote gender equity in international academic medicine. *Medical Teacher*, *40*(9), 962-968.

Inge, S. (2018). Career advice: How to job-share in academia. *Times Higher Education*. https://www.timeshighereducation.com/news/career-advice-how-job-share-academia

Institute for Women's Policy Research & Aspen Institute. (2019). *Parents in College: By the Numbers*. https://iwpr.org/wp-content/uploads/2020/10/Understanding-the-Student-Parent-Experience_Final.pdf

Isasi, F., Naylor, M. D., Skorton, D., Grabowski, D. C., Hernández, S., & Rice, V. M. (2021). Patients, Families, and Communities COVID-19 Impact Assessment: Lessons Learned and Compelling Needs. *NAM Perspectives*, *2021*. https://doi.org/10.31478/202111c

Jacobs, J. A., & Winslow, S. E. (2004). The academic life course, time pressures and gender inequality. *Community, Work & Family*, *7*(2), 143-161. https://doi.org/10.1080/1366880042000245443

Jacobs, J. C., Van Houtven, C. H., Laporte, A., & Coyte, P. C. (2016). The impact of informal caregiving intensity on women's retirement in the United States. *Journal of Population Ageing*, *10*(2), 159-180. https://doi.org/10.1007/s12062-016-9154-2

Jagsi, R., Beeland, T. D., Sia, K., Szczygiel, L. A., Allen, M. R., Arora, V. M., Bair-Merritt, M., Bauman, M. D., Bogner, H. R., Daumit, G., Davis, E., Fagerlin, A., Ford, D. E., Gbadegesin, R., Griendling, K., Hartmann, K., Hedayati, S. S., Jackson, R. D., Matulevicius, S., . . . Escobar Alvarez, S. (2022). Doris Duke Charitable Foundation Fund to Retain Clinical Scientists: Innovating support for early-career family caregivers. *Journal of Clinical Investigation*, *132*(23). https://doi.org/10.1172/jci166075

Jagsi, R., Jones, R. D., Griffith, K. A., Brady, K. T., Brown, A. J., Davis, R. D., Drake, A. F., Ford, D., Fraser, V. J., Hartmann, K. E., Hochman, J. S., Girdler, S., Libby, A. M., Mangurian, C., Regensteiner, J. G., Yonkers, K., Escobar-Alvarez, S., & Myers, E. R. (2018). An innovative program to support gender equity and success in academic medicine: Early experiences from the Doris Duke Charitable Foundation's Fund to Retain Clinical Scientists. *Annals of Internal Medicine, 169*(2), 128-130. https://doi.org/10.7326/m17-2676

Jagsi, R., Tarbell, N. J., & Weinstein, D. F. (2007). Becoming a doctor, starting a family — leaves of absence from graduate medical education. *New England Journal of Medicine, 357*(19), 1889-1891. https://doi.org/10.1056/nejmp078163

Jain, S., Neaves, S., Royston, A., Huang, I., & Juengst, S. B. (2022). Breastmilk pumping experiences of physician mothers: Quantitative and qualitative findings from a nationwide survey study. *Journal of General Internal Medicine, 37*(13), 3411-3418. https://doi.org/10.1007/s11606-021-07388-y

Jean, V., Payne, S., & Thompson, R. (2014). Women in STEM: Family-related challenges and initiatives. Gender and the work–family experience (pp. 291-311). Springer. https://doi.org/10.1007/978-3-319-08891-4_15

Johnson & Johnson. (2018). *Pivot From Career Break to Breakthrough With Re-Ignite*. Johnson & Johnson. https://www.careers.jnj.com/re-ignite

Johnson, R. W., & Lo Sasso, A. T. (2000). *The Trade-Off between Hours of Paid Employment and Time Assistance to Elderly Parents at Midlife*. https://www.urban.org/research/publication/trade-between-hours-paid-employment-and-time-assistance-elderly-parents-midlife

Johnson, R. W., Smith, K. E., & Butrica, B. (2023). *Lifetime Employment-Related Costs to Women of Providing Family Care*. Washington DC: Urban Institute Report. https://www.dol.gov/sites/dolgov/files/WB/Mothers-Families-Work/Lifetime-caregiving-costs_508.pdf

Jones, R. D., Miller, J., Vitous, C. A., Krenz, C., Brady, K. T., Brown, A. J., Daumit, G. L., Drake, A. F., Fraser, V. J., & Hartmann, K. E. (2020). From stigma to validation: A qualitative assessment of a novel national program to improve retention of physician-scientists with caregiving responsibilities. *Journal of Women's Health, 29*(12), 1547-1558. https://www.ncbi.nlm.nih.gov/pmc/articles/PMC7864110/pdf/jwh.2019.7999.pdf

Jones, R. D., Miller, J., Vitous, C. A., Krenz, C., Brady, K. T., Brown, A. J., Daumit, G. L., Drake, A. F., Fraser, V. J., Hartmann, K. E., Hochman, J. S., Girdler, S., Libby, A. M., Mangurian, C., Regensteiner, J. G., Yonkers, K., & Jagsi, R. (2019). The most valuable resource is time: Insights from a novel national program to improve retention of physician–scientists with caregiving responsibilities. *Academic Medicine, 94*(11), 1746-1756. https://doi.org/10.1097/acm.0000000000002903

Kachchaf, R., Ko, L., Hodari, A., & Ong, M. (2015). Career–life balance for women of color: Experiences in science and engineering academia. *Journal of Diversity in Higher Education, 8*(3), 175-191. https://doi.org/10.1037/a0039068

Kahn, J. R., García-Manglano, J., & Bianchi, S. M. (2014). The motherhood penalty at midlife: Long-term effects of children on women's careers. *Journal of Marriage and Family, 76*(1), 56-72. https://doi.org/10.1111/jomf.12086

REFERENCES

Kalra, K., Delaney, T. V., & Dagi Glass, L. R. (2023). Perceptions of parental leave among ophthalmologists. *JAMA Ophthalmology*, *141*(1), 24-31. https://doi.org/10.1001/jamaophthalmol.2022.5236

Keene, J. R., & Prokos, A. H. (2007). The sandwiched generation: Multiple caregiving responsibilities and the mismatch between actual and preferred work hours. *Sociological Spectrum*, *27*(4), 365-387.

Kennelly, I., & Spalter-Roth, R. M. (2006). Parents on the job market: Resources and strategies that help sociologists attain tenure-track jobs. *The American Sociologist*, *37*(4), 29-49. https://doi.org/10.1007/bf02915066

Klerman, J. A., Daley, K., & Pozniak, A. (2012). *Family and Medical Leave in 2012: Technical Report.* https://www.dol.gov/sites/dolgov/files/OASP/legacy/files/FMLA-2012-Technical-Report.pdf

Kleven, H., Landais, C., Posch, J., Steinhauer, A., & Zweimüller, J. (2019). Child penalties across countries: Evidence and explanations. *AEA Papers and Proceedings*, *109*, 122-126. https://doi.org/10.1257/pandp.20191078

Kmec, J. A. (2013). Why academic STEM mothers feel they have to work harder than others on the job. *International Journal of Gender, Science and Technology*, *5*(2), 79-101. https://genderandset.open.ac.uk/index.php/genderandset/article/view/277

Knoll, M. A., Griffith, K. A., Jones, R. D., & Jagsi, R. (2019). Association of gender and parenthood with conference attendance among early career oncologists. *JAMA Oncology*, *5*(10), 1503-1504. https://doi.org/10.1001/jamaoncol.2019.1864

Kokorelias, K. M., Gignac, M. A. M., Naglie, G., & Cameron, J. I. (2019). Towards a universal model of family centered care: A scoping review. *BMC Health Services Research*, *19*(1). https://doi.org/10.1186/s12913-019-4394-5

Koppes Bryan, L., & Wilson, C. A. (2015). *Shaping work-life culture in higher education.* Routledge. https://doi.org/10.4324/9780203118887

Kossek, E., Lautsch, B., & Eaton, S. (2009). "Good teleworking": Under what conditions does teleworking enhance employees' well-being? *Technological and Psychological Well-being*. Cambridge University Press. https://doi.org/10.1017/CBO9780511635373.007

Kossek, E., Lewis, S., & Hammer, L. B. (2010). Work–life initiatives and organizational change: Overcoming mixed messages to move from the margin to the mainstream. *Human Relations*, *63*(1), 3-19. https://doi.org/10.1177/0018726709352385

Kossek, E., Pichler, S., Bodner, T., & Hammer, L. (2011). Workplace social support and work-family conflict: A meta-analysis clarifying the influence of general and work-family-specific supervisor and organizational support. *Personnel Psychology*, *64*, 289-313. https://doi.org/10.1111/j.1744-6570.2011.01211.x

Kossek, E. E. (2006). Work and family in America: Growing tensions between employment policy and a transformed workforce. In E. E. Lawler & J. O'Toole (Eds.), *America at work: Choices and challenges* (pp. 53-71). Palgrave Macmillan US. https://doi.org/10.1057/9781403983596_4

Kossek, E. E., Dumas, T. L., Piszczek, M. M., & Allen, T. D. (2021). Pushing the boundaries: A qualitative study of how STEM women adapted to disrupted work–nonwork boundaries during the COVID-19 pandemic. *Journal of Applied Psychology*, *106*(11), 1615-1629. https://doi.org/10.1037/apl0000982

Kossek, E. E., Lautsch, B. A., Perrigino, M. B., Greenhaus, J. H., & Merriweather, T. J. (2023). Work-life flexibility policies: Moving from traditional views toward work-life intersectionality considerations. *Research in Personnel and Human Resources Management*, 41. https://doi.org/10.1108/s0742-730120230000041008

Kossek, E. E., & Lee, K.-H. (2020). *Fostering Gender and Work-Life Inclusion for Faculty in Understudied Contexts: An Organizational Science Lens*. Purdue e-Pubs. https://docs.lib.purdue.edu/worklifeinclusion/worklifeinclusion_monograph.pdf

Kossek, E. E., & Lee, K.-H. (2022). Work-life inclusion for women's career equality: Why it matters and what to do about it. *Organizational Dynamics*, *51*(2), 100818. https://doi.org/https://doi.org/10.1016/j.orgdyn.2020.100818

Kossek, E. E., Perrigino, M., & Rock, A. G. (2021). From ideal workers to ideal work for all: A 50-year review integrating careers and work-family research with a future research agenda. *Journal of Vocational Behavior*, *126*. https://doi.org/10.1016/j.jvb.2020.103504

Kossek, E. E., Perrigino, M. B., & Lautsch, B. A. (2023). Work-life flexibility policies from a boundary control and implementation perspective: A review and research framework. *Journal of Management*, 01492063221140354.

Kossek, E. E., Thompson, R. J., Lawson, K. M., Bodner, T., Perrigino, M. B., Hammer, L. B., Buxton, O. M., Almeida, D. M., Moen, P., & Hurtado, D. A. (2019). Caring for the elderly at work and home: Can a randomized organizational intervention improve psychological health? *Journal of Occupational Health Psychology*, *24*(1), 36. https://www.ncbi.nlm.nih.gov/pmc/articles/PMC5991990/pdf/nihms909657.pdf

Kotsadam, A., & Finseraas, H. (2011). The state intervenes in the battle of the sexes: Causal effects of paternity leave. *Social Science Research*, *40*(6), 1611-1622. https://doi.org/https://doi.org/10.1016/j.ssresearch.2011.06.011

Kraus, M. B., Talbott, J. M., Melikian, R., Merrill, S. A., Stonnington, C. M., Hayes, S. N., Files, J. A., & Kouloumberis, P. E. (2021). Current parental leave policies for medical students at US medical schools: A comparative study. *Academic Medicine*, *96*(9), 1315-1318. https://www.ingentaconnect.com/content/wk/acm/2021/00000096/00000009/art00038;jsessionid=aapcl8bqqk1ri.x-ic-live-01

Kupp, H. (2021). Measuring success in a digital-first world. https://futureforum.com/2021/10/05/how-to-measure-outcomes-playbook/

Ladores, S., Debiasi, L., & Currie, E. (2019). Breastfeeding women in academia: Pursuing tenure track versus "mommy" track. *Clinical Lactation*, *10*, 11-16. https://doi.org/10.1891/2158-0782.10.1.11

Lambert, W. M., Nana, N., Afonja, S., Saeed, A., Amado, A. C., & Golightly, L. M. (2022). *Addressing structural mentoring barriers in postdoctoral training: A qualitative study*. Cold Spring Harbor Laboratory. https://dx.doi.org/10.1101/2022.08.20.504665

LaPonsie, M. (2022). How much does it cost to raise a child? *US News*. https://money.usnews.com/money/personal-finance/articles/how-much-does-it-cost-to-raise-a-child

Lawson, K. M., Barrineau, M., Woodling, C. M., Ruggles, S., & Largent, D. L. (2023). The impact of COVID-19 on U.S. computer science faculty's turnover intentions: The role of gender. *Sex Roles*, *88*(7-8), 383-396. https://doi.org/10.1007/s11199-023-01361-1

Lee, J., Williams, J. C., & Li, S. (2017). *Parents in the Pipeline: Retaining Postdoctoral Researchers with Families*. https://repository.uclawsf.edu/cgi/viewcontent.cgi?article=1001&context=wll

Lee, Y., & Tang, F. (2015). More caregiving, less working: Caregiving roles and gender difference. *Journal of Applied Gerontology*, *34*(4), 465-483. https://doi.org/10.1177/0733464813508649

Leibnitz, G., & Morrison, B. K. (2015). The eldercare crisis and implications for women faculty. In *Disrupting the culture of silence* (pp. 137-145). Routledge.

Lerner, D. J. (2022). Invisible overtime: What employers need to know about caregivers. In *Family Caregivers and Employment White Paper*. Rosalynn Carter Institute for Caregivers.

Levine, R. B., Walling, A., Chatterjee, A., & Skarupski, K. A. (2022). Factors influencing retirement decisions of senior faculty at U.S. medical schools: Are there gender-based differences? *Journal of Women's Health*, *31*(7), 974-982. https://doi.org/10.1089/jwh.2021.0536

Lino, M., Kuczynski, K., Rodriguez, N., & Schap, T. (2017). *Expenditures on Children by Families, 2015*. https://fns-prod.azureedge.us/sites/default/files/resource-files/crc2015-march2017.pdf

Lubitow, A., & Zippel, K. (2014). Strategies of academic parents to manage work-life conflict in research abroad. *Gender Transformation in the Academy*, *19*, 63-84. https://doi.org/10.1108/S1529-212620140000019003

Lundquist, J., Misra, J., & O'Meara, K. (2012). Parental leave usage by fathers and mothers at an American university. *Fathering: A Journal of Theory Research and Practice about Men as Fathers*, *10*, 337-363. https://doi.org/10.3149/fth.1003.337

MacCormick, H. (2015). *Stanford's "Time banking" Program helps emergency room Physicians Avoid Burnout*. https://web.archive.org/web/20230324221117/https://scopeblog.stanford.edu/2015/08/21/stanfords-time-banking-program-helps-emergency-room-physicians-avoid-burnout/

Mackenzie, A., & Greenwood, N. (2012). Positive experiences of caregiving in stroke: A systematic review. *Disability and Rehabilitation*, *34*(17), 1413-1422. https://doi.org/10.3109/09638288.2011.650307

Maddi, A., Monneau, E., Guaspare, C., Gargiulo, F., & Dubois, M. (2023). *PubPeer and Self-Correction of Science: Male-Led Publications More Prone to Retraction*. https://hal.science/hal-04246117

Maestas, N., Messel, M., & Truskinovsky, Y. (2023). *Caregiving and Labor Supply: New Evidence from Administrative Data*. https://dx.doi.org/10.3386/w31450

Magudia, K., Bick, A., Cohen, J., Ng, T. S. C., Weinstein, D., Mangurian, C., & Jagsi, R. (2018). Childbearing and family leave policies for resident physicians at top training institutions. *JAMA*, *320*(22), 2372. https://doi.org/10.1001/jama.2018.14414

Manchester, C. F., Leslie, L. M., & Kramer, A. (2013). Is the clock still ticking? An evaluation of the consequences of stopping the tenure clock. *ILR Review*, *66*(1), 3-31. https://doi.org/10.1177/001979391306600101

Marcus, J. (2007). Helping academics have families and tenure too: Universities discover their self-interest. *Change: The Magazine of Higher Learning*, *39*(2), 27-32. https://doi.org/10.3200/CHNG.39.2.27-32

Marcus, J. (2019). Most Americans don't realize state funding for higher ed fell by billions. *Public Broadcasting Service (PBS)*. https://www.pbs.org/newshour/education/most-americans-dont-realize-state-funding-for-higher-ed-fell-by-billions

Marotte, M. R., Reynolds, P., & Savarese, R. J. (2011). *Papa, PhD: Essays on fatherhood by men in the academy*. Rutgers University Press.

Mason, C. N. (2022). *All in Together: The Role of the Family Resource Center in Achieving Student Parent Success at Los Angeles Valley College*. https://iwpr.org/wp-content/uploads/2022/08/LAVC-Brief-All-In-Together-FINAL.pdf

Mason, M. A., Goulden, M., & Frasch, K. (2007). *Graduate student parents: The underserved minority*. Council of Graduate Schools, Washington, DC.

Mason, M. A., Wolfinger, N. H., & Goulden, M. (2019). *Do babies matter? Gender and family in the ivory tower*. Rutgers University Press.

Mason, M. A., & Younger, J. (2014). Title IX and pregnancy discrimination in higher education: The new frontier. *N.Y.U. Review of Law & Social Change*, *38*, 269.

Matulevicius, S. A., Kho, K. A., Reisch, J., & Yin, H. (2021). Academic medicine faculty perceptions of work-life balance before and since the COVID-19 pandemic. *JAMA Network Open*, *4*(6), e2113539. https://doi.org/10.1001/jamanetworkopen.2021.13539

Mayo Clinic. (2023). *Job burnout: How to spot it and take action*. https://www.mayoclinic.org/healthy-lifestyle/adult-health/in-depth/burnout/art-20046642

Mazanec, S. R., Daly, B. J., Douglas, S. L., & Lipson, A. R. (2011). Work productivity and health of informal caregivers of persons with advanced cancer. *Research in Nursing & Health*, *34*(6), 483-495. https://doi.org/10.1002/nur.20461

McAlear, F., Scott, A., Scott, K., & Weiss, S. (2018). *Data Brief: Women and Girls of Color in Computing*. Kapor Center, ASU Center for Gender Equity in Science and Technology, Pivotal Ventures. https://www.wocincomputing.org/wp-content/uploads/2018/08/WOCinComputingDataBrief.pdf

McCann, J. J., Hebert, L. E., Beckett, L. A., Morris, M. C., Scherr, P. A., & Evans, D. A. (2000). Comparison of informal caregiving by black and white older adults in a community population. *Journal of the American Geriatrics Society*, *48*(12), 1612-1617. https://doi.org/10.1111/j.1532-5415.2000.tb03872.x

Melin, J. L. (2023). The help-seeking paradox: Gender and the consequences of using career reentry assistance. *Social Psychology Quarterly*, *0*(0), 01902725231180804. https://doi.org/10.1177/01902725231180804

Mensah, C., Engel, L. O., Bolger, D., & Lee, J. (2022). *Final Title IX NPRM Comment from Pregnant and Parenting Students Advocates*. The Federal Advocacy Coalition for Pregnant and Parenting Students Members.

Miller, A. R. (2009). The effects of motherhood timing on career path. *Journal of Population Economics*, *24*(3), 1071-1100. https://doi.org/10.1007/s00148-009-0296-x

Misra, J., Kuvaeva, A., O'Meara, K., Culpepper, D. K., & Jaeger, A. (2021). Gendered and racialized perceptions of faculty workloads. *Gender & Society*, *35*(3), 358-394. https://doi.org/10.1177/08912432211001387

Misra, J., Lundquist, J. H., & Templer, A. (2012). Gender, work time, and care responsibilities among faculty. *Sociological Forum*, *27*(2), 300-323. https://doi.org/10.1111/j.1573-7861.2012.01319.x

Moon, H. E., Haley, W. E., Rote, S. M., & Sears, J. S. (2020). Caregiver well-being and burden: Variations by race/ethnicity and care recipient nativity status. *Innovation in Aging*, *4*(6), igaa045. https://doi.org/10.1093/geroni/igaa045

Moore, M. R. (2017). Women of color in the academy: Navigating multiple intersections and multiple hierarchies. *Social Problems*, *64*(2), 200-205. https://doi.org/10.1093/socpro/spx009

Moors, A., Stewart, A., & Malley, J. (2022). Managing the career effects of discrimination and motherhood: The role of collegial support for a caregiving policy at a research-intensive U.S. university. *Journal of Higher Education Policy and Management*, 1-16. https://doi.org/10.1080/1360080x.2022.2076188

Morain, S., Schoen, L., Marty, M., & Schwarz, E. B. (2019). Parental leave, lactation, and childcare policies at top US schools of public health. *American Journal of Public Health*, *109*(5), 722-728. https://doi.org/10.2105/ajph.2019.304970

Morgan, A. C., Way, S. F., Hoefer, M. J. D., Larremore, D. B., Galesic, M., & Clauset, A. (2021). The unequal impact of parenthood in academia. *Science Advances*, *7*(9), eabd1996. https://doi.org/doi:10.1126/sciadv.abd1996

Morris, L., Calvert, C. T., & Lee, J. (2021). *Litigation or Clarification? The Impact of Family Responsibilities Discrimination Laws*. https://worklifelaw.org/wp-content/uploads/2021/07/Litigation-or-Clarification-The-Impact-of-Family-Responsibilities-Discrimination-Laws.pdf

Myers, B. (2018). *The Power of Giving Caregivers Extra Hands for Their Research*. Doris Duke Charitable Foundation. https://www.dorisduke.org/news--insights/insights/power-giving-caregivers-extra-hands-their-research/

National Academies of Sciences, Engineering, and Medicine. (2016). *Families caring for an aging America*. The National Academies Press. https://doi.org/10.17226/23606

National Academies of Sciences, Engineering, and Medicine. (2019). *Taking action against clinician burnout: A systems approach to professional well-being*. The National Academies Press. https://doi.org/10.17226/25521

National Academies of Sciences, Engineering, and Medicine. (2020). *Promising Practices for addressing the underrepresentation of women in science, engineering, and medicine: Opening doors*. The National Academies Press. https://doi.org/10.17226/25585

National Academies of Sciences, Engineering, and Medicine. (2021). *The impact of COVID-19 on the careers of women in academic sciences, engineering, and medicine*. The National Academies Press. https://doi.org/10.17226/26061

National Academies of Sciences, Engineering, and Medicine. (2023). *Barriers, challenges, and supports for family caregivers in science, engineering, and medicine: Proceedings of two symposia*. The National Academies Press. https://doi.org/10.17226/27181

National Academy of Medicine. (2022). *Action collaborative on clinician well-being and resilience*. https://nam.edu/initiatives/clinician-resilience-and-well-being/

National Aeronautics and Space Administration. (2021). *Balancing Your Research with Your Life*. National Aeronautics and Space Administration. https://science.nasa.gov/researchers/work-life-balance

National Center for Science and Engineering Statistics. (2021). *Diversity and STEM: Women, Minorities, and Persons with Disabilities 2023*. https://ncses.nsf.gov/pubs/nsf23315/report/

National Institutes of Health. (2021). *Family-friendly initiatives*. National Institutes of Health. https://grants.nih.gov/grants/policy/nih-family-friendly-initiative.htm

National Institutes of Health. (2023). *Research supplements to promote reentry and reintegration into health related research careers*. https://orwh.od.nih.gov/career-development-education/research-supplements-promote-reentry-and-reintegration-health-related

National Science Foundation. (2015). *Chapter II - Proposal Preparation Instructions*. National Science Foundation. https://www.nsf.gov/pubs/policydocs/pappg22_1/pappg_2.jsp#IIE8

Newport, C. (2016). *Deep work: Rules for focused success in a distracted world.* Hachette UK.

Nizalova, O. (2012). The wage elasticity of informal care supply: Evidence from the Health and Retirement Study. *Southern Economic Journal, 79*(2), 350-366. http://www.jstor.org/stable/41638879

Noll, E., Reichlin, L., & Gault, B. (2017). *College Students with Children: National and Regional Profiles* (#C451). https://files.eric.ed.gov/fulltext/ED612519.pdf

O'Brien, K. R., Martinez, L. R., Ruggs, E. N., Rinehart, J., & Hebl, M. R. (2015). Policies that make a difference: Bridging the gender equity and work-family gap in academia. *Gender in Management, 30*(5), 414-426. https://doi.org/https://doi.org/10.1108/GM-02-2014-0013

Office for Civil Rights. (2022). Federal Register Notice of Proposed Rulemaking: Title IX of the Education Amendments of 1972. https://www2.ed.gov/about/offices/list/ocr/docs/t9nprm.pdf

Oldach, L. (2022). High-affinity binding. *ASBMB Today.* https://www.asbmb.org/asbmb-today/people/083122/high-affinity-binding

Ollilainen, M. (2019). Academic mothers as ideal workers in the USA and Finland. *Equality, Diversity and Inclusion: An International Journal, 38*(4), 417-429.

O'Meara, K. (2016). Whose problem is it? Gender differences in faculty thinking about campus service. *Teachers College Record: The Voice of Scholarship in Education, 118*(8), 1-38. https://doi.org/10.1177/016146811611800808

Organisation for Economic Co-operation and Development. (2020). PF2.3: Additional leave entitlements for working parents. https://www.oecd.org/els/soc/PF2_3_Additional_leave_entitlements_of_working_parents.pdf

Organisation for Economic Co-operation and Development. (2022). PF2.1. Parental leave systems. https://www.oecd.org/els/soc/PF2_1_Parental_leave_systems.pdf

Ortiz Worthington, R., Feld, L. D., & Volerman, A. (2019). Supporting new physicians and new parents: A call to create a standard parental leave policy for residents. *Academic Medicine, 94*(11), 1654-1657. https://doi.org/10.1097/acm.0000000000002862

Pang, A. S.-K. (2016). *Rest: Why you get more done when you work less.* Basic books. https://books.google.com/books?hl=en&lr=&id=byjXCwAAQBAJ&oi=fnd&pg=PR7&ots=sJJ9UV7pIT&sig=9_Lj1JXlR8CLAhoLFg3RTlmLP1c#v=onepage&q&f=false

Park, S. S. (2020). Caregivers' mental health and somatic symptoms during COVID-19. *The Journals of Gerontology: Series B, 76*(4), e235-e240. https://doi.org/10.1093/geronb/gbaa121

Patterson, S. E., & Margolis, R. (2019). The demography of multigenerational caregiving: A critical aspect of the gendered life course. *Socius: Sociological Research for a Dynamic World, 5*, 237802311986273. https://doi.org/10.1177/2378023119862737

Peters, C. E., & Hartigan, S. M. (2023). Pregnancy and parental leave in medicine and academia—a focus on urology. *Nature Reviews Urology*, 1-2.

Pierret, C. R. (2006). The sandwich generation: Women caring for parents and children. *Monthly Labor Review, 129*, 3.

Pinquart, M., & Sörensen, S. (2005). Ethnic differences in stressors, resources, and psychological outcomes of family caregiving: A meta-analysis. *The Gerontologist*, 45(1), 90-106. https://doi.org/10.1093/geront/45.1.90

Prados, M., & Zamarro, G. (2020). Gender Differences in couples' division of childcare, work and mental health during COVID-19. *SSRN Electronic Journal*. https://doi.org/10.2139/ssrn.3667803

The Pregnant Scholar. (2020). *Model Campus Lactation Policy for Students*. https://thepregnantscholar.org/wp-content/uploads/Model-Lactation-Policy-for-Students.pdf

The Pregnant Scholar. (2022). *Title IX Communications Guide: Supporting Pregnant and Parenting Students*. https://thepregnantscholar.org/wp-content/uploads/Communications-Guide-for-Campus-Administrators-Fall-2022.pdf

The Pregnant Scholar. (2023). *Supporting pregnant and parenting students: Ideas for faculty*. Center for WorkLife Law.

Quinn, C., Clare, L., & Woods, R. T. (2010). The impact of motivations and meanings on the wellbeing of caregivers of people with dementia: A systematic review. *International Psychogeriatrics*, 22(1), 43-55. https://doi.org/10.1017/S1041610209990810

Raja, U., Chowdhury, N. S., Raje, R. R., Wheeler, R., Williams, J., & Ganci, A. (2021). COVID CV: A System for Creating Holistic Academic CVs during a Global Pandemic. 2021 IEEE International Conference on Electro Information Technology (EIT).

Reddick, R. J., Rochlen, A. B., Grasso, J. R., Reilly, E. D., & Spikes, D. D. (2012). Academic fathers pursuing tenure: A qualitative study of work-family conflict, coping strategies, and departmental culture. *Psychology of Men & Masculinity*, 13(1), 1.

Reichlin-Cruse, L., Richburg-Hayes, L., Hare, A., & Contreras-Mendez, S. (2021). *Evaluating the role of campus child care in student parent success*. Institute for Women's Policy Research.

Reinhard, S. C., Caldera, S., Houser, A., & Choula, R. B. (2023). Valuing the invaluable: 2023 update strengthening supports for family caregivers. *AARP Public Policy Institute*. https://www.aarp.org/ppi/info-2015/valuing-the-invaluable-2015-update.html

Reinhard, S. C., Feinberg, L. F., Houser, A., Choula, R., & Evans, M. (2019). Valuing the invaluable 2019 update: Charting a path forward. *AARP Public Policy Institute*, 146, 1-32.

Reuter, A. A. (2006). Subtle but pervasive: Discrimination against mothers & pregnant women in the workplace. *Fordham Urban Law Journal*, 33(5). https://ir.lawnet.fordham.edu/ulj/vol33/iss5/3/

Riano, N. S., Linos, E., Accurso, E. C., Sung, D., Linos, E., Simard, J. F., & Mangurian, C. (2018). Paid family and childbearing leave policies at top US medical schools. *JAMA*, 319(6), 611. https://doi.org/10.1001/jama.2017.19519

Richmond, G. (2020). *Did the pandemic set women scientists back permanently?* National Science Foundation. https://new.nsf.gov/science-matters/did-pandemic-set-women-scientists-back-permanently

Rinaldo, R., & Whalen, I. M. (2023). Amplifying inequalities: Gendered perceptions of work flexibility and the division of household labor during the COVID-19 pandemic. *Gender, Work & Organization*, 30(6), 1922-1940. https://doi.org/10.1111/gwao.13026

Robertson, A. S., & Weiner, A. (2013). Building community for student-parents and their families: A social justice challenge for higher education. *Journal of Academic Perspectives*. https://www.journalofacademicperspectives.com/app/download/969932205/robertson.pdf

Roselin, D., Lee, J., Jagsi, R., Blair-Loy, M., Ira, K., Dahiya, P., Williams, J., & Mangurian, C. (2022). Medical student parental leave policies at U.S. medical schools. *Journal of Women's Health*, *31*(10), 1403-1410. https://doi.org/10.1089/jwh.2022.0048

Ruppanner, L., Moller, S., & Sayer, L. (2019). Expensive childcare and short school days = lower maternal employment and more time in childcare? Evidence from the American Time Use Survey. *Socius: Sociological Research for a Dynamic World*, *5*, 237802311986027. https://doi.org/10.1177/2378023119860277

Ryan, R. A., Whipps, M. D., & Bihuniak, J. D. (2021). Barriers and facilitators to expressing milk on campus as a breastfeeding student. *Journal of American College Health*, 1-7.

Sacks, J., Valin, S., Casson, R. I., & Wilson, C. R. (2015). Are 2 heads better than 1? Perspectives on job sharing in academic family medicine. *Canadian Family Physician*, *61*(1), 11-13, e11-13.

Sallee, M., & Lester, J. (2009). The family-friendly campus in the 21st century. In *Establishing the family-friendly campus* (pp. 159-165). Routledge.

Sallee, M., Ward, K., & Wolf-Wendel, L. (2016). Can anyone have it all? Gendered views on parenting and academic careers. *Innovative Higher Education*, *41*(3), 187-202. https://doi.org/10.1007/s10755-015-9345-4

Sallee, M. W. (2008). A feminist perspective on parental leave policies. *Innovative Higher Education*, *32*, 181-194.

Sallee, M. W. (2012). The ideal worker or the ideal father: Organizational structures and culture in the gendered university. *Research in Higher Education*, *53*(7), 782-802. https://doi.org/10.1007/s11162-012-9256-5

Samus, Q. M., Johnston, D., Black, B. S., Hess, E., Lyman, C., Vavilikolanu, A., Pollutra, J., Leoutsakos, J.-M., Gitlin, L. N., Rabins, P. V., & Lyketsos, C. G. (2014). A multidimensional home-based care coordination intervention for elders with memory disorders: The Maximizing Independence at Home (MIND) pilot randomized trial. *The American Journal of Geriatric Psychiatry*, *22*(4), 398-414. https://doi.org/10.1016/j.jagp.2013.12.175

Sattari, M., Levine, D. M., Mramba, L. K., Pina, M., Raukas, R., Rouw, E., & Serwint, J. R. (2020). Physician mothers and breastfeeding: A cross-sectional survey. *Breastfeeding Medicine*, *15*(5), 312-320. https://doi.org/10.1089/bfm.2019.0193

Sayer, L. C. (2005). Gender, time and inequality: Trends in women's and men's paid work, unpaid work and free time. *Social Forces*, *84*(1), 285-303. https://doi.org/10.1353/sof.2005.0126

Sayer, L. C., Bianchi, S. M., & Robinson, J. P. (2004). Are parents investing less in children? Trends in mothers' and fathers' time with children. *American Journal of Sociology*, *110*(1), 1-43. https://doi.org/10.1086/386270

Schimpf, C., Santiago, M. M., Hoegh, J., Banerjee, D., & Pawley, A. (2013). STEM faculty and parental leave: Understanding an institution's policy within a national policy context through structuration theory. *International Journal of Gender, Science and Technology*, *5*(2), 102-125.

Schmitt, L., & Auspurg, K. (2022). A stall only on the surface? Working hours and the persistence of the gender wage gap in Western Germany 1985–2014. *European Sociological Review*, *38*(5), 754-769. https://doi.org/10.1093/esr/jcac001

REFERENCES

Schneider, M. C., Graham, L., Hornstein, A. S., LaRiviere, K. J., Muldoon, K. M., Shepherd, S. L., & Wagner, R. (2021). Caregiving, disability, and gender in academia in the time of COVID-19. *Advance Journal, 2*(3).

Shanafelt, T. D., Boone, S., Tan, L., Dyrbye, L. N., Sotile, W., Satele, D., West, C. P., Sloan, J., & Oreskovich, M. R. (2012). Burnout and satisfaction with work-life balance among US physicians relative to the general US population. *Archives of Internal Medicine, 172*(18), 1377-1385. https://doi.org/10.1001/archinternmed.2012.3199

Shauman, K., Howell, L. P., Paterniti, D. A., Beckett, L. A., & Villablanca, A. C. (2018). Barriers to career flexibility in academic medicine: A qualitative analysis of reasons for the under-utilization of family-friendly policies, and implications for institutional change and department chair leadership. *Academic Medicine: Journal of the Association of American Medical Colleges, 93*(2), 246. https://www.ncbi.nlm.nih.gov/pmc/articles/PMC5788717/pdf/nihms890376.pdf

Sidik, S. (2023). Toxic workplaces are the main reason women leave academic jobs. *Nature, 623*(7985), 19. https://doi.org/10.1038/d41586-023-03251-8

Skarupski, K. A., Roth, D. L., & Durso, S. C. (2021). Prevalence of caregiving and high caregiving strain among late-career medical school faculty members: Workforce, policy, and faculty development implications. *Human Resources for Health, 19*(1). https://doi.org/10.1186/s12960-021-00582-3

Skinner, M., Betancourt, N., & Wolff-Eisenberg, C. (2021). The disproportionate impact of the pandemic on women and caregivers in academia. *Ithaka S+R*. https://doi.org/10.18665/sr.315147

Skira, M. M. (2015). Dynamic wage and employment effects of elder parent care. *International Economic Review, 56*(1), 63-93. http://www.jstor.org/stable/24517888

Smith, J. L., Vidler, L. L., & Moses, M. S. (2022). The "gift" of time: Documenting faculty decisions to stop the tenure clock during a pandemic. *Innovative Higher Education, 47*(5), 875-893. https://doi.org/10.1007/s10755-022-09603-y

Smith, P. R. (2004). Elder care, gender, and work: The work-family issue of the 21st century. *Berkeley Journal of Employment and Labor Law, 25*, 351.

Sodders, M. D., Killien, E. Y., Stansbury, L. G., Vavilala, M. S., & Moore, M. (2020). Race/ethnicity and informal caregiver burden after traumatic brain injury: A scoping study. *Health Equity, 4*(1), 304-315. https://doi.org/10.1089/heq.2020.0007

Soffer, M. D. (2019). Current realities of childbearing in residency: Room for improvement. *Obstetrics & Gynecology, 133*(3), 571-574. https://doi.org/10.1097/aog.0000000000003120

Sosa, J. A., & Mangurian, C. (2023). Addressing eldercare to promote gender equity in academic medicine. *JAMA, 330*(23), 2245-2246. https://doi.org/10.1001/jama.2023.23743

Springer, K. W., Parker, B. K., & Leviten-Reid, C. (2009). Making space for graduate student parents: Practice and politics. *Journal of Family Issues, 30*(4), 435-457.

Stack, C. B. (1974). *All our kin: Strategies for survival in a Black community.* New York: Harper & Row.

Stall, N. M., Shah, N. R., & Bhushan, D. (2023). Unpaid family caregiving—the next frontier of gender equity in a postpandemic future. *JAMA Health Forum, 4*(6), e231310. https://doi.org/10.1001/jamahealthforum.2023.1310

STEM Reentry Task Force. (2015). *Return-to-Work Opportunities*. Society of Women Engineers, iRelaunch. https://reentry.swe.org/return-to-work-opportunities/

Stern, S. (1995). Estimating family long-term care decisions in the presence of endogenous child characteristics. *The Journal of Human Resources, 30*(3), 551-580. https://doi.org/10.2307/146035

Stoller, S. E. (2023). *Inventing the working parent: Work, gender, and feminism in neoliberal britain*. MIT Press. https://doi.org/10.7551/mitpress/14918.001.0001

Stoner, J. B., & Stoner, C. R. (2016). Career disruption: The impact of transitioning from a full-time career professional to the primary caregiver of a child with autism spectrum disorder. *Focus on Autism and Other Developmental Disabilities, 31*(2), 104-114. https://doi.org/10.1177/1088357614528798

Strachan, E., & Buchwald, D. (2023). Informal caregiving among American Indians and Alaska Natives in the Pacific Northwest. *Journal of Community Health, 48*(1), 160-165. https://doi.org/10.1007/s10900-022-01156-7

Stygles, K. N. (2016). *Queering Academia: Queer Faculty Mothers and Work-Family Enrichment*. Bowling Green State University. http://rave.ohiolink.edu/etdc/view?acc_num=bgsu1478536020611255

Suh, J. (2016). Measuring the "sandwich": Care for children and adults in the American Time Use Survey 2003–2012. *Journal of Family and Eonomic Issues, 37*, 197-211. https://www.ncbi.nlm.nih.gov/pmc/articles/PMC4883270/pdf/10834_2016_Article_9483.pdf

Super, N. (2002). *Who Will Be There to Care? The Growing Gap Between Caregiver Supply and Demand* [NHPF Background Paper]. National Health Policy Forum. https://hsrc.himmelfarb.gwu.edu/cgi/viewcontent.cgi?article=1082&context=sphhs_centers_nhpf

Swenson, K., & Simms, K. B. (2021). *Increases in Out-Of-Pocket Child Care Costs: 1995 to 2016.* https://aspe.hhs.gov/sites/default/files/migrated_legacy_files/200606/increases-in-out-of-pocket-child-care-costs.pdf

Szczygiel, L. A., Jones, R. D., Drake, A. F., Drake, W. P., Ford, D. E., Hartmann, K. E., Libby, A. M., Marshall, B. A., Regensteiner, J. G., Yaffe, K., & Jagsi, R. (2021). Insights from an intervention to support early career faculty with extraprofessional caregiving responsibilities. *Women's Health Reports, 2*(1), 355-368. https://doi.org/10.1089/whr.2021.0018

Tam, B., Findlay, L., & Kohen, D. (2017). Conceptualization of family: Complexities of defining an Indigenous family. *Indigenous Policy Journal, 28*.

Tay, D. L., Iacob, E., Reblin, M., Cloyes, K. G., Jones, M., Hebdon, M. C. T., Mooney, K., Beck, A. C., & Ellington, L. (2022). What contextual factors account for anxiety and depressed mood in hospice family caregivers? *Psycho-Oncology, 31*(2), 316-325. https://doi.org/10.1002/pon.5816

Templeton, K., Bernstein, C. A., Sukhera, J., Nora, L. M., Newman, C., Burstin, H., Guille, C., Lynn, L., Schwarze, M. L., Sen, S., & Busis, N. (2019). Gender-based differences in burnout: Issues faced by women physicians. *NAM Perspectives*. https://doi.org/10.31478/201905a

Thébaud, S., & Taylor, C. J. (2021). The specter of motherhood: Culture and the production of gendered career aspirations in science and engineering. *Gender & Society, 35*(3), 395-421. https://doi.org/10.1177/08912432211006037

REFERENCES

Thomas, P. (2021). Is a four-day week the future of work? *The Wall Street Journal*. https://www.wsj.com/articles/is-a-four-day-week-the-future-of-work-11627704011?mod=article_inline

Thorndyke, L. E., Cain, J., & Milner, R. J. (2017). *The UMMS/UMMHC Accelerator Plan: Maximizing Options for Faculty to Reach and Sustain their Full Potential*. https://doi.org/10.13028/A233-T891

Torres, I., Collins, N., Hertz, A., & Liukkonen, M. (2023a). *Global Call to Action for Mothers in Science: Action Plan for Funding Agencies 2023*. https://www.mothersinscience.com/action-plan-funding-agencies

Torres, I., Collins, R., Hertz, A., & Liukkonen, M. (2023b). *Policy Proposals to Promote Inclusion of Caregivers in the Research Funding System: A Call for Change*. https://doi.org/10.31235/osf.io/n473p

Tower, L. E., & Latimer, M. (2016). Cumulative disadvantage: Effects of early career childcare issues on faculty research travel. *Affilia*, *31*(3), 317-330. https://doi.org/10.1177/0886109915622527

University of Alabama at Birmingham. (2018). *Sick Time Donation*. University of Alabama at Birmingham. https://www.uab.edu/humanresources/home/records-administration/leave-of-absence/sick-time-donation

University of California, San Francisco. (2020). *Faculty Family-Friendly Initiative (3FI)*. UCSF Office of Faculty and Academic Affairs. https://facultyacademicaffairs.ucsf.edu/faculty-life/3FI

U.S. Bureau of Labor Statistics. (2023a). *American Time Use Survey Summary*. https://www.bls.gov/news.release/atus.nr0.htm

U.S. Bureau of Labor Statistics. (2023b). *Employment Characteristics of Families — 2022*. https://www.bls.gov/news.release/pdf/famee.pdf

U.S. Census Bureau. (2022). *Census Bureau Releases New Estimates on America's Families and Living Arrangements*. https://www.census.gov/newsroom/press-releases/2022/americas-families-and-living-arrangements.html

U.S. Congress. (2021). *H.R.1065 - Pregnant Workers Fairness Act*. https://www.congress.gov/bill/117th-congress/house-bill/1065

U.S. Department of Education. (2023). *Child Care Access Means Parents in School Program*. Office of Postsecondary Education. https://www2.ed.gov/programs/campisp/index.html

U.S. Department of Labor. (1993). *The Family and Medical Leave Act of 1993*. Retrieved from https://www.dol.gov/agencies/whd/laws-and-regulations/laws/fmla

U.S. Department of Labor. (2022). *FLSA Protections to Pump at Work*. Wage and Hour Division. https://www.dol.gov/agencies/whd/pump-at-work

Uttal, L. (1996). Racial safety and cultural maintenance: The childcare concerns of employed mothers of color. *Ethnic Studies Review*, *19*, 43-59. https://doi.org/10.1525/esr.1996.19.1.43

Valantine, H., & Sandborg, C. I. (2013). Changing the culture of academic medicine to eliminate the gender leadership gap: 50/50 by 2020. *Academic Medicine*, *88*(10), 1411-1413. https://doi.org/10.1097/ACM.0b013e3182a34952

Valantine, H. A. (2020). Where are we in bridging the gender leadership gap in academic medicine? *Academic Medicine*, *95*(10), 1475-1476. https://doi.org/10.1097/acm.0000000000003574

Van Houtven, C. H., Coe, N. B., & Skira, M. M. (2013). The effect of informal care on work and wages. *Journal of Health Economics*, *32*(1), 240-252. https://doi.org/https://doi.org/10.1016/j.jhealeco.2012.10.006

Van Osch, Y., & Schaveling, J. (2020). The effects of part-time employment and gender on organizational career growth. *Journal of Career Development*, *47*(3), 328-343. https://doi.org/10.1177/0894845317728359

Vasel, K. (2021). These return-to-work programs could help moms reenter the workforce. *CNN Business*. https://www.cnn.com/2021/06/01/success/returnship-programs/index.html

Vos, E. E., De Bruin, S. R., Van Der Beek, A. J., & Proper, K. I. (2021). "It's like juggling, constantly trying to keep all balls in the air": A qualitative study of the support needs of working caregivers taking care of older adults. *International Journal of Environmental Research and Public Health*, *18*(11), 5701. https://doi.org/10.3390/ijerph18115701

W. S. Badger Company. (2016). *What Happens When You Let Employees Bring Their Babies to Work?* https://www.badgerbalm.com/pages/babies-at-work

Wakabayashi, C., & Donato, K. M. (2005). The consequences of caregiving: Effects on women's employment and earnings. *Population Research and Policy Review*, *24*(5), 467-488. https://doi.org/10.1007/s11113-005-3805-y

Wang, R., & Bianchi, S. M. (2009). ATUS fathers' involvement in childcare. *Social Indicators Research*, *93*(1), 141-145. https://doi.org/10.1007/s11205-008-9387-4

Ward, K., & Wolf-Wendel, L. (2012). *Academic motherhood: How faculty manage work and family*. Rutgers University Press.

Ward, K., & Wolf-Wendel, L. E. (2005). Work and family perspectives from research university faculty. *New Directions for Higher Education*, *2005*(130), 67-80. https://doi.org/10.1002/he.179

Weinstein, D. F., Mangurian, C., & Jagsi, R. (2019). Parenting during graduate medical training-practical policy solutions to promote change. *New England Journal of Medicine*, *381*(11), 995-997.

Weller, C. E., & Tolson, M. (2019). *Unpaid Family Caregiving and Retirement Savings*. Political Economic Research Institute. https://peri.umass.edu/publication/item/1188-unpaid-family-caregiving-and-retirement-savings

Weston, K. (1997). *Families we choose: Lesbians, gays, kinship*. Columbia University Press.

White, C., & Cruse, L. R. (2021). *The Student Parent Equity Imperative: Guidance for the Biden-Harris Administration*. Policy Brief C496. Institute for Women's Policy Research.

Wilcox, V., & Sahni, H. (2022). The effects on labor supply of living with older family members needing assistance with activities of daily living. *Journal of Family and Economic Issues*, *44*(4), 900-918. https://doi.org/10.1007/s10834-022-09880-x

Williams, J. (2000). *Unbending gender: Why family and work conflict and what to do about it*. Oxford University Press.

Williams, J., Boyle, J., Davis, A., Ertman, M., & Polikoff, N. (2000). Unbending gender: Why work and family conflict and what to do about it. *American University Law Review*. https://scholarship.law.bu.edu/cgi/viewcontent.cgi?article=3245&context=faculty_scholarship

Williams, J., & Norton, D. (2010). *Building academic excellence through gender equity*.

REFERENCES

Williams, J. C. (1989). Deconstructing gender. *Michigan Law Review, 87*(4). https://repository.law.umich.edu/mlr?utm_source=repository.law.umich.edu%2Fmlr%2Fvol87%2Fiss4%2F3&utm_medium=PDF&utm_campaign=PDFCoverPages

Williams, J. C. (2005). The glass ceiling and the maternal wall in academia. *New Directions for Higher Education, 2005*(130), 91-105. https://doi.org/10.1002/he.181

Williams, J. C. (2014). Double Jeopardy? An Empirical Study with Implications for the Debates over Implicit Bias and Intersectionality. *Harvard Journal of Law and Gender.* https://nwlc.org/wp-content/uploads/2021/03/Double-Jeopardy_-An-Empirical-Study-with-Implications-for-the-Deb.pdf

Williams, J. C., Berdahl, J. L., & Vandello, J. A. (2016). Beyond work-life "integration". *Annual Review of Psychology, 67*(1), 515-539. https://doi.org/10.1146/annurev-psych-122414-033710

Williams, J. C., Korn, R. M., & Ghani, A. (2022). *Pinning Down the Jellyfish: The Workplace Experiences of Women of Color in Tech.* https://worklifelaw.org/publication/pinning-down-the-jellyfish-the-workplace-experiences-of-women-of-color-in-tech/

Williams, J. C., Korn, R. M., & Mihaylo, S. (2020). Beyond implicit bias: Litigating race and gender employment discrimination using data from the Workplace Experiences Survey. *Hastings Law Journal, 72*(1). https://repository.uchastings.edu/hastings_law_journal/vol72/iss1/7

Williams, J. C., & Lee, J. (2016). Is it time to stop stopping the clock. *The Chronicle of Higher Education.* https://www.chronicle.com/article/is-it-time-to-stop-stopping-the-clock/

Williams, J. C., Multhaup, M., Li, S., & Korn, R. (2018). *You Can't Change What You Can't See: Interrupting Racial & Gender Bias in the Legal Profession.* https://mcca.com/wp-content/uploads/2018/09/You-Cant-Change-What-You-Cant-See-Executive-Summary.pdf

Wilson, R. (2008). More colleges offer part-time options for professors. *The Chronicle of Higher Education, 54*(45), B26. https://www.chronicle.com/article/more-colleges-offer-part-time-options-for-professors/

Winslow, S., & Davis, S. (2016). Gender inequality across the academic life course. *Sociology Compass, 10,* 404-416. https://doi.org/10.1111/soc4.12372

Witters, D. (2011). Caregiving Costs U.S. Economy $25.2 Billion in Lost Productivity. *Gallup.* https://news.gallup.com/poll/148670/caregiving-costs-economy-billion-lost-productivity.aspx

Wladis, C., Hachey, A. C., & Conway, K. (2023). Time poverty: A hidden factor connecting online enrollment and college outcomes? *The Journal of Higher Education, 94*(5), 609-637.

Wolfinger, N. H., Mason, M. A., & Goulden, M. (2009). Stay in the game: Gender, family formation and alternative trajectories in the academic life course. *Social Forces, 87*(3), 1591-1621. http://www.jstor.org/stable/40345173

WorkLife Law. (2017). *Family Caregiver Discrimination.* https://worklifelaw.org/projects/family-caregiver-discrimination

World Health Organization. (2019). Burn-out an "occupational phenomenon": International Classification of Diseases. https://www.who.int/news/item/28-05-2019-burn-out-an-occupational-phenomenon-international-classification-of-diseases

Worthman, C. M., Plotsky, P. M., Schechter, D. S., & Cummings, C. A. (2010). *Formative experiences: The interaction of caregiving, culture, and developmental psychobiology.* Cambridge University Press.

Writer, J. H., & Watson, D. C. (2019). Recruitment and retention: An institutional imperative told through the storied lenses of faculty of color. *Journal of the Professoriate, 10*(2), 23-46. https://caarpweb.org/wp-content/uploads/2019/11/Haynes-Writer-Watson-Recruitment-and-Retention.pdf

Wynn, A. T. (2019). Change without an agent: What happens when change agents leave? *Organizational Dynamics, 48*(4), 100723. https://doi.org/https://doi.org/10.1016/j.orgdyn.2019.04.009

Xu, L., Tang, F., Li, L. W., & Dong, X. Q. (2017). Grandparent caregiving and psychological well-being among Chinese American older adults—the roles of caregiving burden and pressure. *The Journals of Gerontology: Series A, 72*(Suppl 1), S56-S62. https://doi.org/10.1093/gerona/glw186

Xu, Y. J., & Martin, C. L. (2011). Gender differences in STEM disciplines: From the aspects of informal professional networking and faculty career development. *Gender Issues, 28*(3), 134-154. https://doi.org/10.1007/s12147-011-9104-5

Zamarro, G., & Prados, M. J. (2021). Gender differences in couples' division of childcare, work and mental health during COVID-19. *Review of Economics of the Household, 19*(1), 11-40. https://doi.org/10.1007/s11150-020-09534-7

Zanhour, M., & Sumpter, D. M. (2022). The entrenchment of the ideal worker norm during the COVID-19 pandemic: Evidence from working mothers in the United States. *Gender, Work & Organization.* https://doi.org/10.1111/gwao.12885

Zheng, X., Yuan, H., & Ni, C. (2022). How parenthood contributes to gender gaps in academia. *eLife, 11*, e78909. https://doi.org/10.7554/eLife.78909

Zippel, K. S. (2017). *Women in global science: Advancing academic careers through international collaboration.* Standford, CA: Stanford University Press.

Appendix A

Biographical Sketches of Committee Members and Commissioned Paper Authors

COMMITTEE

Elena Fuentes-Afflick (*Chair*) is professor of pediatrics and vice dean for the UCSF School of Medicine at Zuckerberg San Francisco General Hospital at the University of California, San Francisco (UCSF). Throughout her career, Dr. Fuentes-Afflick has personally managed and mentored faculty and staff on a range of caregiving issues in the context of academic medicine. In 2010, Dr. Fuentes-Afflick was elected to membership in the National Academy of Medicine (NAM) and has served on numerous consensus committees, the Membership Committee, and the Diversity Committee; she was elected to the Governing Council and the Executive Committee of Council, and was elected Home Secretary. In 2020, Dr. Fuentes-Afflick was elected to the American Academy of Arts and Sciences. Dr. Fuentes-Afflick obtained her undergraduate and medical degrees at the University of Michigan and a master's degree in public health (epidemiology) from the University of California, Berkeley. She completed her pediatric residency and chief residency at UCSF, followed by a research fellowship at the Phillip R. Lee Institute for Health Policy Studies at UCSF.

Marianne Bertrand is the Chris P. Dialynas Distinguished Service Professor of Economics at the University of Chicago Booth School of Business. Born in Belgium, Professor Bertrand received a bachelor's degree in economics from Belgium's Université Libre de Bruxelles in 1991, followed by a master's degree in econometrics from the same institution the next year. She earned a Ph.D. in economics from Harvard University in 1998. She was a faculty member in the Department of Economics at Princeton University for 2 years before joining Chicago Booth in 2000. Professor Bertrand is an applied micro-economist whose research covers the fields of labor economics, corporate finance, political economy, and development economics. Her research in these areas has been published widely, including numerous research articles in the *Quarterly Journal of Economics*, the *Journal of Political Economy*, the *American Economic Review*, and the *Review of Economic Studies*. Professor Bertrand is a co-director of Chicago Booth's Rustandy Center for Social Sector Innovation and the director of the Inclusive Economy Lab at the University of Chicago Urban Labs. Professor Bertrand also served as co-editor of the *American Economic Review*. She has received several awards and honors, including the 2004 Elaine Bennett Research Prize, awarded by the American Economic Association to recognize and honor outstanding research in any field of economics by a woman at the beginning of her career, and the 2012 Society of Labor Economists' Rosen Prize for Outstanding Contributions to Labor Economics. Professor Bertrand is a research fellow at the National Bureau of Economic Research, the Center for Economic Policy Research, and the Institute for the Study of Labor. She is also a fellow of the American Academy of Arts and Sciences and of the Econometric Society, and a member of the National Academy of Sciences.

Mary Blair-Loy is professor of sociology at the University of California, San Diego. She uses multiple methods to study gender, work, and family. Much scholarship emphasizes individuals' strategic trade-offs or implicit biases. In contrast, Professor Blair-Loy analyzes normative cultural models of a worthwhile life, including the "work devotion schema" (which defines professional work as a calling that penalizes involved caregiving) and the "schema of scientific excellence" (which defines scholarly excellence in terms of culturally masculine traits such as assertive self-promotion). Her 2022 book *Misconceiving Merit: Paradoxes of Excellence and Devotion in Academic Science and Engineering* with Erin Cech uses multiple types of evidence to show that these cultural schemas are broadly embraced yet harm scientists and science. A 2022 article in *Science* shows how hiring rubrics can devalue women academic engineers. A 2019 *Proceedings of the National Academy of Sciences* (*PNAS*) article with Cech uses longitudinal data to show substantial attrition of new mothers from science, technology, engineering, and mathematics (STEM) and was recognized as a Top 10 *PNAS* Article of 2019 to make a "large impact on the public understanding of science." Professor Blair-Loy has been recognized as a Top Ten Extraordinary Contributor in the Landmark Contributions category in the international field of work-family research. She holds a B.A. and Ph.D. from the University of Chicago and an M.Div. from Harvard.

Kathleen Christensen founded the Alfred P. Sloan Foundation's Workplace, Workforce, and Working Families program in 1994 and spearheaded it for two decades. Under her strategic leadership, the foundation has been credited as a driving force in creating the work-family research field and with launching the first national, multicollaborator movement to make workplace flexibility a compelling national issue and the standard of the American workplace. She designed the Faculty Career Flexibility Awards program that, with the American Council on Education, recognized more than 40 colleges, universities, and medical schools for their innovative policies and practices. The endowed Kathleen Christensen Dissertation Award was established by Society of Human Resources Managers and the Work and Family Researchers Network to encourage doctoral candidates and early-career scholars to achieve high and rigorous standards in work-family research. She has been honored as one of the Top Ten Extraordinary Contributors to Work and Family Research (2018), one of the Seven Wonders of the Work-Life Field by *Working Mother* magazine (2010), and with the inaugural Work Life Legacy Award by the Families and Work Institute (2004). Dr. Christensen planned and participated in the 2010 White House Forum on Workplace Flexibility, as well as the 2014 White House Summit on Working Families. She is the recipient of Danforth, Mellon, and National Endowment for the Humanities fellowships. She has authored/edited seven books, including some of the earliest research on working at home and on contingent work. Prior to the Sloan Foundation, Dr. Christensen was a professor of psychology at the Graduate Center of City University of New York. She currently serves as a faculty fellow at Boston College's Center for Social Innovation, where she co-directs Work Equity, a new initiative to address inequities that are institutionalized employment systems. She received her Ph.D. from Pennsylvania State University in geography and philosophy of science.

J. Nicholas Dionne-Odom is an associate professor in the School of Nursing at the University of Alabama at Birmingham (UAB) and co-director of Caregiver and Bereavement Support Services in the UAB Center for

Palliative and Supportive Care. Dr. Dionne-Odom is board certified in hospice and palliative care advanced practice nursing with more than 10 years clinical experience in critical care and 10 years in telehealth palliative care coaching. He is a nationally and internationally recognized expert in developing and testing early palliative interventions for family caregivers of individuals with serious illness, focusing particularly on historically under-resourced populations. Dr. Dionne-Odom's research has totaled $9 million from the National Institute of Nursing Research, the National Cancer Institute, the National Palliative Care Research Center, the Gordon and Betty Moore Foundation, the Cambia Health Foundation, Sigma Theta Tau International, the American Association of Critical Care Nursing, and the UAB Center for Palliative and Supportive Care. In 2020, he received the Protégé Award from the Friends of the National Institute of Nursing Research and was inducted as a fellow in the American Academy of Nursing. Dr. Dionne-Odom acquired his B.S.N. degree from Florida State University (2002), an M.A. in philosophy and education from Teachers College, Columbia University (2006), an M.S.N. in nursing at Boston College (2010), and his Ph.D. in nursing at Boston College (2013).

Mignon Duffy is a professor of sociology at the University of Massachusetts Lowell. Her primary research interests center on care work—the work (paid and unpaid) of taking care of others, including children and those who are elderly, ill, or disabled. She is particularly interested in how the social organization of care intersects with gender, race, class, and other systems of inequality. Her most recent project is an edited volume (co-edited with Amy Armenia and Kim Price-Glynn) that is forthcoming from Rutgers University Press entitled *From Crisis to Catastrophe: Care, COVID, and Pathways to Change*. She is also the co-editor of *Caring on the Clock: The Complexities and Contradictions of Paid Care* (2015) and the author of *Making Care Count: One Hundred Years of Gender, Race, and Paid Care Work* (2011). Dr. Duffy is also a longtime leader (currently serving as past chair) of the Carework Network, an international organization of care work researchers and advocates. Her research has appeared in peer-reviewed journals such as *Gender & Society* and *Social Problems*. Committed to connecting her research to policy, Dr. Duffy has worked in collaboration with policy organizations such as the United Nations, the International Labor Organization, and the World Economic Forum.

Jeff Gillis-Davis is a professor of physics at Washington University in St. Louis. Previously, he was faculty at the University of Hawai'i at Mānoa

(2003–2018). Dr. Gillis-Davis combines experiments, remote sensing, and sample analysis to study the geology of the Moon, Mercury, and asteroids. His primary research area centers on a process known as space weathering. To study space weathering in the laboratory, he uses lasers to replicate the impact of dust-sized particles on the surfaces of these airless bodies. The intense impact energy of these dust-sized particles transforms minerals into glass, can destroy polar ice deposits, or leads to intriguing chemical processes. Dr. Gillis-Davis leads a team of researchers who study the complex processes and environments that determine where ice will be, how it may be modified, how water was delivered to the Moon, and its active water cycle. This team is called the Interdisciplinary Consortium for Evaluating Volatile Origins, or ICE Five-O, one of NASA's Solar System Exploration Research Virtual Institute, or SSERVI. He has also participated as a science team member in three NASA missions: Clementine, Lunar Reconnaissance Orbiter Miniature Radio-Frequency team, and MESSENGER.

Reshma Jagsi is chair of the Department of Radiation Oncology at Emory University and Winship Cancer Institute. A graduate of Harvard College, Harvard Medical School, and the University of Oxford, where she studied as a British Marshall Scholar, she completed her residency training and an ethics fellowship at Harvard before joining the faculty of the University of Michigan, where she served as the director of its Center for Bioethics and Social Sciences in Medicine. Gender equity in academic medicine has been a key area of her scholarly focus, a subject to which she brings her perspective as a physician and social scientist, to promote evidence-based intervention. Author of more than 400 articles in peer-reviewed journals, including multiple high-impact studies in journals such as the *New England Journal of Medicine*, the *Lancet*, and *JAMA*, her research to promote gender equity has been funded by R01 grants from the National Institutes of Health (NIH) as well as large independent grants from the Doris Duke Foundation and several other philanthropic foundations. Her Doris Duke Foundation grant has focused specifically on the development and evaluation of programs intended to support academic medical faculty with family caregiving demands, including an initiative that began well before the outbreak of the COVID-19 pandemic and a new program inspired by the pandemic and the National Academies of Sciences, Engineering, and Medicine report on COVID-19 and women. She has mentored dozens of others in research investigating women's underrepresentation in senior positions in academic medicine and the mechanisms that must be targeted to promote equity.

Active in organized medicine, she has served on the Steering Committee of the Association of American Medical Colleges' (AAMC's) Group on Women in Medicine in Science. She now serves on the National Academies' Committee on Women in Science, Engineering, and Medicine and the Advisory Committee for Research on Women's Health for the NIH. She was part of the *Lancet*'s advisory committee for its theme issue on women in science, medicine, and global health, which served to foster additional research. An internationally recognized clinical trialist and health services researcher in breast cancer, her work is frequently featured in the popular media, including coverage by the *New York Times*, *Wall Street Journal*, and NPR. Her contributions have been recognized with her election to the American Society for Clinical Investigation and Association of American Physicians, the Leadership Award of the AAMC's Group on Women in Medicine and Science, LEAD Oncology's Woman of the Year Award, American Medical Women's Association's Woman in Science Award, and American Medical Student Association's Women Leaders in Medicine Award. She is a fellow of the American Society of Clinical Oncology, American Society for Radiation Oncology, American Association for Women in Radiology, American Association for the Advancement of Science, and the Hastings Center.

Ellen Ernst Kossek is the Basil S. Turner Distinguished Professor in the Krannert School of Management at Purdue University. Prior to joining Purdue, Dr. Kossek was University Distinguished Professor at Michigan State University. She is the first elected president of the Work and Family Researchers Network, and has won dozens of awards for research and service excellence related to advancing the organizational work and family research stream in the field of management. She is an internationally recognized researcher who studies how employment policies and practices to support positive work-family-life relationships impact gender equality and diversity, equity, and inclusion. She designs and conducts field experiments to help organizations and leaders implement work-life flexibility, and work-life cultural change and gender and diversity equality initiatives. Dr. Kossek is elected a fellow in the Academy of Management, the Society for Industrial and Organizational Psychology, and the American Psychological Association. She holds a Ph.D. in organizational behavior from the Yale School of Management, an M.B.A. from the University of Michigan, and a B.A. with honors in psychology from Mount Holyoke College. She led in writing a report for the National Academy of Sciences on the effects of COVID-19 on the work-life boundaries and domestic labor of women in academic STEMM.

Lindsey Malcom-Piqueux is the assistant vice president for diversity, equity, inclusion, and assessment and the chief institutional research officer at the California Institute of Technology (Caltech). In this role, she develops and implements research-informed, metrics-driven institutional efforts to ensure that Caltech is a diverse, equitable, and inclusive environment for all community members. She also oversees all areas of institutional research in support of the institute's planning and decision-making processes. Her scholarly research focuses on understanding the institutional conditions that advance racial and gender equity in STEM fields. Prior to joining Caltech, she served as the associate director of research and policy at the Center for Urban Education at the University of Southern California (USC) and was a research associate professor in the USC Rossier School of Education. She has also held faculty positions at the George Washington University and the University of California, Riverside. Dr. Malcom-Piqueux earned her Ph.D. in urban education with an emphasis in higher education from the University of Southern California, her M.S. in planetary science from Caltech, and her S.B. in planetary science from the Massachusetts Institute of Technology. She has previously served on the National Academies' study committees Increasing Diversity and Inclusion in the Leadership of Competed Space Missions and Developing Indicators for Undergraduate STEM Education.

Sandra Kazahn Masur is a basic scientist and an activist for women in science and medicine at the Icahn School of Medicine at Mount Sinai in New York where she is professor of ophthalmology and of pharmacological sciences and director of its Office for Women's Careers within the Office for Gender Equity in Science and Medicine. Her NIH-funded research explored the cell biology of membrane transport and of corneal wound healing. In active support of scientists, she chaired the Women in Cell Biology Committee of the American Society for Cell Biology (ASCB) and was co-director of the National Eye Institute's Fundamental Issues in Vision Research at the Marine Biology Laboratory, Woods Hole, Massachusetts. She was a participant in the NIH Office for Research in Women's Health strategic planning for Women in Science. The Sandra K. Masur Senior Leadership Award was established by the ASCB to honor individuals with exemplary achievements in cell biology who are also outstanding mentors. She received the Women in Medicine Silver Achievement Award from the Association of American Medical Colleges and the Outstanding Woman Scientist award of the Association for Women in Science and is an elected fellow of the ASCB.

Maria (Mia) Ong is a senior research scientist at TERC, a research and development organization dedicated to STEM education that is based in Cambridge, Massachusetts. Prior to working at TERC, Dr. Ong served on faculty at Swarthmore College, Wellesley College, and Harvard University Graduate School of Education. For nearly three decades, she has researched the experiences of women of color and members of other marginalized groups in computer science, engineering, physics, and general STEM higher education and professions, with emphases on qualitative studies and literature synthesis projects. She has led or co-led numerous projects funded by the U.S. Department of Education, the National Institutes of Health, and the National Science Foundation (NSF). She holds more than 40 solo- or co-authored publications on equity and inclusion topics, including career-life balance, caregiving, counterspaces, and changing cultural norms in STEM. Dr. Ong has served on several national committees and task forces, including the NSF Committee on Equal Opportunities in Science and Engineering (2008–2014), the Social Science Advisory Board of the National Center for Women & Information Technology (2008–2022; chair 2017–2018), the American Institute of Physics National Task Force to Elevate African American Representation in Physics & Astronomy (TEAM-UP, 2017–2020), and the National Academies Committee on Addressing the Underrepresentation of Women of Color in Tech (2019–2022). She is a co-recipient of a Presidential Award for Excellence in Science, Mathematics, and Engineering Mentoring (1998) and a co-recipient of the Excellence in Physics Education Award from the American Physical Society (2022). Dr. Ong holds a Ph.D. in social and cultural studies in education from the University of California, Berkeley.

Robert L. Phillips, Jr., is the founding executive director of the Center for Professionalism and Value in Health Care in Washington, D.C. He is a practicing family physician with training in health services and primary care research. His research seeks to inform clinical care and policies that support it. He leads a national primary care registry with related research on social determinants of health, rural health, and changes in primary care practice. Dr. Phillips has often served the U.S. Department of Health and Human Services, including as vice chair of the Council on Graduate Medical Education, co-chair of the Population Health Subcommittee of the National Committee on Vital and Health Statistics, and on the Negotiated Rule-Making Committee on Shortage Designation. Dr. Phillips was elected to the NAM in 2010, and he was a Fulbright Specialist to the Netherlands and New Zealand. Dr. Phillips completed medical school at the University of Florida, where he graduated with honors for special distinction, and trained clinically in family medicine at the University of Missouri, where he completed a National Research Service Award Fellowship. He currently serves as the chair of the NAM Membership Committee and has served on multiple consensus studies, contributed to several workshops, and served as a reviewer.

Jason Resendez is the president and CEO of the National Alliance for Caregiving (NAC), where he leads research, policy, and innovation initiatives to build health, wealth, and equity for America's 53 million family caregivers. Mr. Resendez is a nationally recognized expert on family care, aging, and the science of inclusion in research. In 2020, he was named one of America's top influencers in aging by PBS's *Next Avenue* alongside Michael J. Fox and Surgeon General Dr. Vivek Murthy. Prior to joining NAC, he was the founding executive director of the UsAgainstAlzheimer's Center for Brain Health Equity and was the principal investigator of a Healthy Brain Initiative cooperative agreement with the Centers for Disease Control and Prevention. While at UsAgainstAlzheimer's, he pioneered the concept of brain health equity through peer-reviewed research, public health partnerships, and public policy. He has been quoted by the *Washington Post*, the *Wall Street Journal*, *STAT News*, *Time*, *Newsweek*, and Univision on health equity issues and has received the Service Award for Caregiving from the National Hispanic Council on Aging; the LULAC Presidential Medal of Honor; and the HerMANO Award from MANA, A National Latina Organization, for his advocacy on behalf of the Latino community.

Hannah Valantine received her M.B.B.S. degree from London University, cardiology fellowship at Stanford, and doctor of medicine from London University. She was appointed assistant professor of medicine, rising to full professor of medicine in 2000, and becoming the inaugural senior associate dean for diversity and leadership in 2004. She pursued a data-driven transformative approach to this work, receiving the NIH Director's Pathfinder Award. Dr. Francis Collins, NIH director, recruited her in 2014 as the inaugural NIH chief officer for scientific workforce diversity, and as a tenured investigator in the National Heart, Lung, and Blood Institute's intramural research program, where she established the laboratory of transplantation genomics. Dr. Valantine is a nationally recognized pioneer in her field, with more than 200 peer-reviewed publications, patents, and sustained NIH funding. She was elected to the National Academy of Medicine in 2020 for her pioneering research in organ transplantation and in workforce diversity.

Joan Williams is the Sullivan Professor of Law and founding director of the Center for WorkLife Law at UC College of the Law, San Francisco. Professor Williams has played a central role in reshaping the conversation about work, gender, and class over the past quarter century. Her path-breaking work helped create the field of work-family studies and modern workplace flexibility policies. She is one of the most cited scholars in her field and is the author of 11 books and more than 100 academic articles. Her many honors include President's Award, Society of Women Engineers (2019); Top Ten Extraordinary Contributor to Work and Family Research Award, Work and Family Researchers Network (2018); Work Life Legacy Award, Families and Work Institute (2014); and Outstanding Scholar Award, Fellows of the American Bar Foundation (2012). Professor Williams received her J.D. from Harvard Law School and an M.A. from the Massachusetts Institute of Technology.

COMMISSIONED PAPER AUTHORS

Ngoc Dao is an assistant professor at the College of Business and Public Management at Kean University, and a research fellow at the Center for Financial Security (CFS) Retirement and Disability Research Center (RDRC), University of Wisconsin–Madison. Dr. Dao studies economics of retirement and caregiving, with a particular focus on the relationship between public policies and retirement behaviors, health utilization of, and caregiving for older adults and disabled individuals. One of her works studies the impact of public policies (including Medicaid, Earned Income Tax Credit, and State Family Paid Leave) on formal and informal (family) caregiving provisions. Her research has been funded by the National Institutes of Health and CFS RDRC at the University of Wisconsin–Madison.

Erin Frawley is an education equity program manager at the Center for WorkLife Law. Prior to joining WorkLife Law, she worked with the San Francisco Unified School District, building the capacity of school site staff to work toward anti-racism and authentic partnership with all students and families. She has also taught English as a second language, worked as a reading comprehension specialist, and developed workshops and curricula for several nonprofit organizations and schools. Within the realm of education, Ms. Frawley is most passionate about large-scale systems change informed by student voice and working to bridge the divide between research, theory, and practice to support educational equity. She holds a bachelor's degree in English literature and psychology from the University of Connecticut and a master's degree in human development and psychology from Harvard's Graduate School of Education.

Jessica Lee is a senior staff attorney at the Center for WorkLife Law, and director of the center's Pregnant Scholar Initiative, the nationwide legal resource center for pregnant and parenting students. Her research and advocacy advances gender and racial equity in the workplace and in education, and she is a nationally recognized expert on the laws at the intersection of employment, education, and maternal and infant health. She provides a wide scope of partner organizations with know-your-rights training and strategic tools. Model legislation co-drafted by Ms. Lee has been introduced in Congress, and at the state level she regularly advises state and local enforcement agencies, and she has guided dozens of educational institutions through drafting and implementing family-responsive policies. She also provides know-your-rights resources and trainings to educate parents and change-makers on the legal rights of caregivers in the workplace and in education—translating complicated legal issues into approachable and useful tools for thousands of nonlawyers. During the COVID-19 crisis, she used her expertise to advance pandemic-related policies to support parents and other caregivers, and she manages the center's free legal helpline. Ms. Lee's work has been covered by a variety of press, from the *New York Times* to the BBC, and her writing has appeared in publications ranging from *Harvard Business Review* and the *Chronicle of Higher Education* to law reviews and medical journals.

Ashley Lowe is a researcher on the Transformative Research Unit for Equity at RTI International. She has a decade of experience conducting intervention and evaluation studies related to community, youth, and sexual violence, racial and community justice, mental health promotion, and the prevention of substance abuse. Ms. Lowe leads mixed-methods research projects, including tasks related to monitoring and evaluation, quantitative and qualitative data analysis, and survey development. Her goal is to support individuals living in communities disproportionately impacted by violence through individual, community, and system-level interventions.

Jennifer Lundquist is associate dean of the College of Social and Behavioral Sciences and professor of sociology at the University of Massachusetts Amherst. A social demographer with an emphasis on race and ethnic stratification, family formation patterns, and immigration, Dr. Lundquist evaluates racial disparities along a variety of demographic outcomes, including marriage, family stability, fertility, and health. Her work in this area extends to an exploration of the neighborhood effects of residential segregation as well as a reevaluation of race relations from a social contact hypothesis perspective. Recent work includes the 2021 book (with Celeste Curington and Ken-Hou Lin) *The Dating Divide: Race and Intimacy in the Era of Online Romance*. She has published her research in a variety of journals, including *Social Forces*, *American Journal of Sociology*, and *American Sociological Review*, and is the co-author of a well-known demography textbook. Her research has been funded by the National Science Foundation, Mellon Foundation, and Alexander von Humboldt Foundation and covered by media outlets including *Time*, *Newsweek*, the *Washington Post*, the *New York Times*, and National Public Radio.

Tasseli McKay is a postdoctoral research fellow in the Department of Sociology at Duke University, working with mentor Christopher Wildeman. She is also affiliated with RTI's Transformative Research Unit for Equity. Dr. McKay serves as principal investigator on the Institutional Contact and Family Violence in an Era of Mass Incarceration project funded by the Eunice Kennedy Shriver National Institute of Child Health and Human Development. She holds a doctorate in social policy from the London School of Economics and Political Science, an M.P.H. from the University of North Carolina at Chapel Hill, and a B.A. in American studies from Yale University. Previously, Dr. McKay worked on the Multi-site Family Study of Incarceration, Parenting, and Partnering, a mixed-methods longitudinal study of two thousand families affected by incarceration. This culminated in her first book, *Holding On: Family and Fatherhood During Incarceration and Reentry* (University of California Press, 2019) with Megan Comfort, Christine Lindquist, and Anupa Bir. Dr. McKay's most recent book, *Stolen Wealth, Hidden Power: The Case for Reparations for Mass Incarceration* (University of California Press, 2022), finds that the steep direct costs of mass-scale imprisonment are far overshadowed by its hidden costs and harms, many of which have been kept out of sight by women's labor. She argues that reparations to Black Americans are critical to any effort to bring mass incarceration to an end.

APPENDIX A

Joya Misra is the Provost Professor of Sociology and Public Policy and Roy J. Zuckerberg Endowed Leadership Chair and previously served as director of the Institute for Social Science Research (ISSR) at the University of Massachusetts Amherst (UMass), from 2019 to 2023. She is also the president of the American Sociological Association and a recent Samuel F. Conti Faculty Fellowship Award winner. Dr. Misra is deeply committed to a publicly engaged social science, with the aim of leveraging knowledge to foster more equitable societies. Her research and teaching primarily focus on social inequality, including inequalities by gender and gender identity, race, ethnicity, sexuality, nationality, citizenship, parenthood status, and educational level. Her work explores the role that policies play in both mediating and entrenching inequality, and primarily falls into the subfields of race/gender/class, political sociology, work and labor, family, and welfare states. Five major grants from the National Science Foundation since 2006 have helped to support Dr. Misra's research, which in 2009 garnered her and Michelle Budig the inaugural World Bank/Luxembourg Income Study Gender Research Award. She is currently principal investigator on the 5-year, $3 million NSF Advance Institutional Transformation grant, sustaining ISSR's leadership role in this important vehicle for advancing equity at the intersections of race and gender in science careers at UMass. Other accolades include the SBS Outstanding Teaching Award (2004–2005), UMass Sociology Mentoring Award (2009–2010, 2014–2015) and Sociologists for Women in Society Mentoring Award (2010). She served as editor of *Gender & Society*, one of the most important journals in the fields of sociology and gender studies.

Joanna Riccitelli is a Ph.D. student in sociology at the University of Massachusetts Amherst and the assistant editor of the *American Sociological Review*. Her research focuses on gender, health, well-being, and higher education. Her current project explores the public discourse around the HPV (human papillomavirus) vaccine in the United States.

Monica Sheppard is the co-director of the Transformative Research Unit for Equity (TRUE) Emerging Equity Scholar (EES) Program, which provides holistic mentorship through well-being support and professional development. In this role Ms. Sheppard is responsible for directing the well-being component of the mentoring program, training, and supporting the Wellbeing Mentors, co-creating the EES Curriculum, and participating in other essential program functions such as the EES Speaker Series and Soft

Skills sessions. Aside from EES, Ms. Sheppard's current projects include work as the qualitative data collection lead in a project aimed to reduce racial and ethnic disparities in the criminal justice system of a Pennsylvania county, and a site liaison and assistant task lead for a research project in support of advancing pretrial policy and research. In addition to project work, Ms. Sheppard is involved in several initiatives aimed at addressing and improving RTI's mission for equity, diversity, inclusion, and belonging. Her positions as the EES well-being director and TRUE's well-being liaison has positioned her to collaborate on future research studies that center well-being, and strategy sessions with different groups at RTI to think through how best to support RTI employee well-being.

Sarah Stoller is a freelance writer, editor, and research consultant. She received her Ph.D. in history from the University of California, Berkeley. Her writing on labor, feminism, and parenthood, as well as popular culture and the crisis of higher education, has appeared in both public and scholarly venues, including *Slate*, *Salon*, the *Los Angeles Review of Books*, *Aeon Magazine*, *Jezebel*, the *Washington Post*, and more. Her first book, *Inventing the Working Parent: Work, Gender, and Feminism in Neoliberal Britain*, was published in 2023 with MIT Press.

Courtney Van Houtven is a professor in the Department of Population Health Science, Duke University School of Medicine, and Duke-Margolis Center for Health Policy. She is also a research scientist in Health Services Research and Development in Primary Care at the Durham Veteran's Administration. Dr. Van Houtven's aging and economics research interests encompass long-term care financing, intrahousehold decision-making, informal care, and end-of-life care. She examines how family caregiving affects health care utilization, expenditures, health and work outcomes of care recipients and caregivers. She directs the VA-Cares Evaluation Center in the Durham COIN (Center of Innovation), which recently completed a national evaluation of the Department of Veterans Affairs' Program of Comprehensive Assistance for Family Caregivers, a program that supports family members who care for injured post-9/11 veterans. She is a co-investigator on the NIA/NIH CARE IDEAS R01 study examining outcomes among care partners and persons with cognitive impairment and dementia and on an R01 called Informal Caregiver Burden in Advanced Cancer: Economic and Health Outcomes.

Appendix B

Caregiver Interview Sample and Methodology

INTERVIEW STUDY METHODS

Study methods were designed to surface the experiences of family caregivers working in the academic sciences, engineering, and medicine whose perspectives have been underrepresented in prior research. A qualitative approach was developed to allow space for the complexity (and often, emotion) surrounding subjective experiences of managing caregiving and career, the multilayered contexts in which those experiences occur, and their consequences for caregivers and the field at large (Sofaer, 1999). The study research protocol and all outreach and data collection materials were reviewed by the National Academies of Sciences, Engineering, and Medicine Institutional Review Board.

SAMPLING

To ensure substantial representation of women of color and caregivers of intersecting marginalized identities, the Committee on Women in Science, Engineering, and Medicine (CWSEM) staff, study committee members, and the RTI International study team partnered to develop a targeted outreach campaign for the study. RTI and CWSEM staff focused on identifying and connecting with member listservs and similar communication tools that centered scholars of color; first-generation college graduates; immigrant scholars; those who identified as lesbian, gay, bisexual, transgender, or queer (LGBTQ+); and those living with disabilities.

Respondents were reached via a series of general and targeted outreach emails and via forwarding to individual contacts. Outreach messages described study eligibility criteria, interviewing approach, and compensation and directed interested individuals to a web-based screening form for the study.

The web-based screening form, developed and hosted by RTI, consisted of a set of closed-ended questions designed to establish study eligibility and elicit information on other relevant life experiences for case selection purposes. To be eligible, respondents had to be working or studying in academic science, engineering, and medical fields (including doctors and nurses in training or academic medicine) and must have had regular, unpaid caregiving responsibilities during or beyond the early COVID-19 pandemic.

RTI reviewed incoming responses daily to identify eligible individuals and select a purposeful sample from among them. Based on the committee's assessment of perspectives that were lacking in prior research, priority was accorded to women and other caregivers of color as well as those who are LGBTQ+, first generation, immigrant, and/or living with disabilities. RTI also sampled for diversity in

- the career path and stage, with the aim of including individuals at all career stages, on and off the tenure track, and with special priority given to women of color in senior faculty roles;
- the nature of the unpaid caregiving responsibilities, including for whom care was given and what kinds of activities the caregiver performed; and
- the science, technology, engineering, mathematics, and medicine (STEMM) discipline, to ensure participation from subfields identified by the study committee as more heavily male dominated.[1]

Outreach, screening, sampling, and recruitment proceeded iteratively based on the evolving characteristics of the available sample. RTI extended the recruitment period by approximately 1 month in order to support ongoing efforts to engage women of color senior faculty.

---

[1] The "heavily male-dominated fields" identified by the study committee were physics; computer science; astronomy or astrophysics; and civil, aerospace, electrical, and mechanical engineering.

DATA COLLECTION

Individuals selected for interviews were contacted using each of their preferred modes of contact (email, phone, or both) and invited to schedule a Zoom interview with an experienced qualitative interviewer. Individuals who completed the screening form but were not selected to participate in an interview were thanked immediately and were notified at the end of the recruitment period, using their preferred mode of contact, that they had not been selected.

An informed consent process was administered at the beginning of each scheduled Zoom interview and only individuals who consented to participate in the interview and to be recorded were interviewed. Interviews lasted approximately 1 hour and followed a semistructured guide that covered the following topics:

- Experiences of managing career and caregiving responsibilities simultaneously
- The macro-, meso- and micro-level contexts in which caregivers managed those demands
- Ideas for reimagining success and productivity
- Experiences of joy and satisfaction in career and caregiving

Following the interview, each respondent was sent a thank-you email that included a $75 Amazon gift code, information about the expected release of study findings, and contact information for the CWSEM representative.

ANALYSIS

Interview recordings were professionally transcribed for analysis. A deductive codebook was developed based on the study research questions and early study committee guidance. Inductive codes were developed jointly by the research team to reflect themes that emerged during the interviews. New inductive codes were added by all members of the research team during the analytic process.

In addition, a set of family codes was developed to reflect differences in life experience that the study committee expected might be meaningful for purposes of managing caregiving and career, including the nature of the respondent's caregiving role(s), the forms of care they provided, career path and stage, ethnic and racial identity, and other experiences of structural disadvantage (identifying as LGBTQ+, an immigrant, a first-generation college graduate, and/or living with a disability).

Recognizing the multitude of distinct life experiences of potential relevance and the complex, intersecting nature of those experiences, the team did not attempt to sort or otherwise "bin" the transcripts according to the family codes (as in a structured, comparative analysis). Rather, analysts referred to screening data to apply family codes to each set of interview notes at the beginning of the transcript review. This information informed the coding and theming process, and analysts made note of any inductive observations related to the family codes.

Themes were described in brief analytic memos shared across the analysis team via a shared master analytic file along with the exemplary quotations associated with each theme.

The interviews were coded with family codes detailing key characteristics of each interviewee as well as deductive codes that emerged from the data. Tables B-1 and B-2 present the codes used in analysis.

TABLE B-1 Family Codes

Domain	Family Codes
Nature of caregiving role(s) (all that apply)	Caring for a young child, under the age of 5
	Caring for a school-age minor child, age 5–18
	Caring for an adult child with disabilities or other health needs
	Caring for a partner or spouse with disabilities or other health needs
	Caring for an elder family member, such as a parent or grandparent
	Caring for a member of extended family
	Caring for a chosen family member, friend, or anyone else
Forms of caregiving (all that apply)	Supporting physical subsistence, such as dressing, bathing, toileting, feeding
	Supporting daily living in nonphysical ways (such as living with a cognitive impairment, developmental disability, or autism)
	Supporting participation in online school or other remote learning
	Coordinating medical or behavioral health care
	Managing day-to-day schedule
	Accompanying and/or transporting to regular appointments
	Managing involvement w/legal system (immigration, JJ, CLS)
	Managing finances
	Other regular, unpaid caregiving responsibilities

TABLE B-1 Continued

Domain	Family Codes
Career changes made to manage caregiving (all that apply or none)	Dropped out of school or out of the paid workforce entirely
	Left an academic STEMM field
	Reduced work hours or switched to part-time status within field
	Opted for more flexible work commitments within your field
	Made other major educational or professional changes
	No major changes/not applicable
Career path	Academic position
	Position outside of academia
Career stage or academic "rank"	Student or trainee (graduate student, resident, postdoc, etc.)
	Junior tenure-track faculty (assistant professor)
	Midcareer tenure-track faculty (tenured associate professor)
	Senior faculty (tenured professor, dean, other leadership position)
	Non-tenure-track academic position (lecturer, senior lecturer, adjunct, temporary, research associate)
Gender context in field	Working in physics, computer science, astronomy and astrophysics, or civil, aerospace, electrical, or mechanical engineering
	Working in another field
Ethnic and racial identity (all that apply)	Hispanic/Latino
	American Indian or Alaska Native
	Asian
	Black or African American
	Native Hawaiian or Other Pacific Islander
	White
Other experiences of structural disadvantage (all that apply)	First-generation college graduate
	Immigrant
	Living with a disability
	LGBTQ+

TABLE B-2 Deductive Codes

Domain	Code
Influences	• Macro-level (e.g., community cultural wealth, structural disadvantage) • Meso-level (e.g., everyday interactions within institutions and social support networks) • Micro-level (e.g., personal identities, priorities, household composition)
Work-life management	• Autonomy (or lack thereof) in caregiving • Autonomy (or lack thereof) in professional life • Career constraints associated with caregiving (e.g., impact on career options or ability to meet expectations of position) • Caregiving constraints associated with career (e.g., limitations on time with children or elders) • Personal consequences of managing joint responsibilities (e.g., stress) • Individual strategies for managing joint responsibilities • Career achievement relative to expectation
Informal supports	• Supportive figures in personal life • Other supportive aspects of personal circumstances • Supportive figures in professional life • Other supportive aspects of professional circumstances (not covered elsewhere)
Access to work-life policies	• Helpful formal supports • Unhelpful formal supports • Formal supports not accessed • Reasons for using/not using formal supports • Impact of using/not using formal supports • Biggest help you could have been offered but were not
Reimagining productivity	• Research-related productivity ideals (e.g., funding) • Dissemination-related productivity ideals (e.g., presentations publications) • Alternative productivity ideals
Reimagining success	• Academic prestige • Tenure process • Alternative markers of success
Satisfaction and joy	• Ways that caregiving supported professional contributions • Greatest source(s) of joy in professional life • Greatest source(s) of joy in caregiving

APPENDIX B   191

TABLE B-2 Continued

Domain	Code
Modifier Codes (any domain)	• Helped (facilitated, supported) • Hindered (constrained, was a barrier) • Lost (diminished, worsened) • Gained (enhanced, strengthened, improved) • Stayed the same (neutral, no effect)

SAMPLE CHARACTERISTICS

Reflecting the study aims and sampling objectives, a majority of interview participants were caregivers of color (N = 26). Roughly one-quarter of the sample identified as Black (N = 9), Latinx (N = 11), or Asian (N = 10); half identified as White (N = 21); and smaller numbers identified as American Indian or Alaska Native (N = 3), and/or Native Hawaiian or Other Pacific Islander (N = 2).

Sixty percent of sample members were from immigrant families. A majority of sample members were the first person in their families to graduate from a 4-year college (N = 21).

Women of color senior faculty (N = 8), caregivers working in heavily male-dominated fields (N = 4), LGBTQ+ caregivers (N = 4), and caregivers with disabilities (N = 7) were represented in smaller but substantial numbers. They were drawn from across all career stages, from students to senior faculty and academic leadership, with heaviest representation from graduate students, medical residents, and other early-career scholars (see Figure B-1).

Career Stage

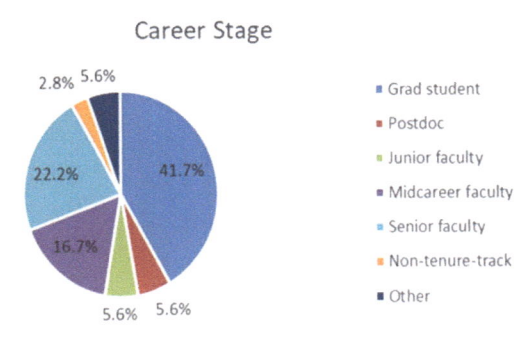

- Grad student: 41.7%
- Postdoc: 5.6%
- Junior faculty: 16.7%
- Midcareer faculty: 22.2%
- Senior faculty: 2.8%
- Non-tenure-track: 5.6%
- Other: 5.6%

FIGURE B-1 Career stages.

A majority of qualitative interview participants were engaged in caregiving for young (35 percent) or school-age (38 percent) children. Substantial proportions, however, were caring for other loved ones: One-third were caring for a parent or other elder family member (33 percent), and smaller proportions caring for chosen family or friends (15 percent), spouses or partners (10 percent), members of extended family (8 percent), or an adult child with disabilities or other intensive caregiving needs (8 percent).

Participants fulfilled a wide variety of caregiving roles. Most provided support with activities of daily living as well as a range of other high- and low-autonomy responsibilities such as supporting physical needs (75 percent), coordinating medical care (73 percent), accompanying or transporting to activities/appointments (65 percent), managing day-to-day scheduling (53 percent), supporting participation in online school (43 percent), managing finances (25 percent), and managing involved with the legal system (3 percent). Study participants reported having made a range of career changes due to the imperatives of their caregiving responsibilities, including opting for more flexible work (60 percent), reducing working hours (40 percent), and leaving their field (15 percent). A small proportion (13 percent) of sample members had not made any major career changes due to their caregiving responsibilities.

REFERENCE

Sofaer, S. 1999. Qualitative methods: What are they and why use them? *Health Services Research, 34*(5 Pt. 2), 1101-1118.

Appendix C

Methodology for Selecting Causal Analyses of the Economic Impact of Caregiving

The literature review, titled "The Economic Impact of Family Caregiving for Women in Academic STEMM: Driving an Evidence-Based Policy Approach,"[1] was conducted from May 1, 2023, to July 14, 2023, using Google Scholar (papers considered from 2000 onward), papers recommended by the National Academies of Sciences, Engineering, and Medicine committee, past evidence syntheses, book chapters on older adult caregiving (co-authored by the team), and conference or working papers identified via abstracts sourced across multiple professional associations over the last 5 years (Allied Social Sciences Association, Population Association of America, American Society of Health Economists, National Bureau of Economic Research, and Association for Public Policy Analysis and Management). Citations located within identified papers were also considered, particularly those sourced from review articles and the most recent literature, including papers emerging post-COVID-19. U.S. studies were first considered due to the context of the U.S. social safety net (e.g., no universal paid sick leave or universal maternity leave, little collective bargaining, privately paid childcare as the norm), as well as international studies. This review focused on work and thriving at work outcomes, shown in Table C-1.

Search terms were used to identify literature regarding older and disabled adult family caregiving (terms: informal car*, family caregiv*, unpaid care*, carer, and work outcomes [see Table C-1]) and child family caregiving

---

[1] The full review is available at https://nap.nationalacademies.org/resource/27416.

TABLE C-1 Work Outcomes

1. Labor force participation (any work)
2. Early retirement/labor market exit
3. Hours of work, part-time, full-time
4. Earnings, wages, wage penalties
5. Reentry into labor force, return to work
6. Job/career opportunities: promotion, tenure or tenure denial, time in rank, moves off the tenure track
7. Occupational status/attainment
8. Productivity at work, work productivity and impairment index, number of publications, citations

(terms: parent*, mother*, father*, having child*, new mother*, birth, parenthood penalty, motherhood penalty, gender disparity, gender gap, and work outcomes [see Table C-1]). Parenthood literature was not limited by the age of children in the home, but instead focused on dependent minor children or disabled adult children in the home. Where possible, child and older or disabled adult caregiving are distinguished and the findings are integrated based on the preponderance of the evidence.

A focus was given to causal studies where possible; causal methods are crucial because mothers may self-select into different occupations and fields where "non-pecuniary benefits related to motherhood are larger" (Simonsen and Skipper, 2006). Causal studies considered as part of this review included laboratory, audit, and quasi-experimental studies. Parenthood timing is also often a choice and depends on many unobserved and observed factors. Nonrandom selection of caregivers for adults may arise in several ways, including the choice of an older adult sibling to enter into caregiving based on the opportunity and time cost for each sibling, comparatively. Identified high-quality studies are noted, such as those presenting longitudinal data or contributing unique data, but that are correlational. Noncausal studies that may produce correlations are stated as such.

REFERENCE

Simonsen, M., & Skipper, L. 2006. The costs of motherhood: An analysis using matching estimators. *Journal of Applied Econometrics, 21*(7), 919-934. http://www.jstor.org/stable/25146477

Appendix D

Current Federal, State, and Local Policies to Support Family Caregivers

This appendix draws substantially from the research paper "Comprehensive Literature Review of Current and Promising Practices to Support Unpaid Caregivers in Science, Technology, Engineering, Mathematics, and Medicine (STEMM)," by Jessica Lee, J.D., Erin Frawley, M.Ed., and Sarah Stoller, Ph.D., which was commissioned for this study.[1]

CAREGIVING LEAVE

Leave is a cornerstone of supporting family caregivers in science, technology, engineering, mathematics, and medicine (STEMM). Accessible leave—leave that caregivers can take without significant adverse career, educational, or financial repercussions—results in more equitable parenting throughout the lifespan and may ameliorate the negative impacts of caregiving on employee mental health and productivity (Heshmati et al., 2023).

FAMILY AND MEDICAL LEAVE ACT

The federal Family and Medical Leave Act (FMLA) requires covered employers to provide their eligible employees with unpaid, job-protected leave for up to 12 weeks in a 12-month period (U.S. Department of Labor, 1993). During the leave period, employers must continue to provide their employees with continued health insurance coverage. This law applies to all public institutions as well as private employers with more than 50 employees.

Employees eligible for FMLA leave must have worked at least 1,250 hours for the employer in the 12 months prior to the start of leave, must have worked for the employer for at least 12 months total, and must work

---

[1] The full paper is available at https://nap.nationalacademies.org/resource/27416.

in a location with at least 50 employees in a 75-mile radius (U.S. Department of Labor, 1993). Eligible employees can use the leave for pregnancy, bonding with a newborn or newly adopted/placed child, or to care for a member of the employee's immediate family with a serious health condition ("immediate family" is defined as child, spouse, employee's own parent). The FMLA defines "serious health conditions" to include physical or mental health conditions requiring an overnight stay in hospital or similar facility; conditions that incapacitate the family member for more than 3 consecutive days and require ongoing medical treatment such as follow-up appointments and/or medication; chronic conditions that incapacitate and require treatment at least twice a year; or pregnancy.

The FMLA has special considerations for employee spouses working for the same employer. These employees are limited in the amount of time they can take for bonding with a newborn or newly adopted/placed child. Leave for child bonding is limited to 12 weeks total for dual-career employees at the same institution, while leave for one's own pregnancy, health condition, or child's health condition is not split. For example, if a faculty member takes 8 weeks of pregnancy FMLA leave and 4 weeks for baby bonding, those 4 weeks could be deducted from the 12 that her faculty spouse is eligible to use for baby bonding (U.S. Department of Labor, 2023).

To access this leave, employees should provide 30-day advanced notice when the need is foreseeable and offering notice is practicable. If the need for leave is not foreseeable, the employee should provide notice as soon as practicable. Employees may be required to fill out paperwork confirming their relative's medical condition. Following the employee's leave, they must be reinstated to their job or one that is nearly identical. Leave may be taken intermittently, but employers have a choice of whether to allow intermittent leave to bond with a newborn or newly placed child.

Finally, the FMLA prohibits employers from interfering with an employee's ability to take leave or retaliating against them for taking leave. The law is enforced through the U.S. Department of Labor and private lawsuits.

STATE AND LOCAL LEAVE LAWS

At least 16 states provide their own job-protected leave for caregiving employees (National Conference of State Legislatures, 2015). These laws are typically very similar to the federal FMLA, though they often have expanded eligibility, such as by reducing employer size thresholds or the length of time an employee must have worked to be eligible for leave. Twelve states and Washington, D.C., have a law requiring paid leave for new parents and family caregivers (A Better Balance, 2023). Several of these laws have been recently enacted and are not yet providing benefits. Notably, state paid family leave laws typically have caps on benefit amounts (e.g., no more than $900 a week) and as such are typically unable to fully replace a faculty member's pay.

TITLE IX

Title IX of the Education Amendments of 1972 (Title IX) prohibits discrimination on the basis of sex and requires educational institutions to provide their students/trainees and employees with leave related to pregnancy.

Students and nonemployee trainees must be provided leave for pregnancy and related conditions (such as childbirth or miscarriage recovery) for as long as deemed medically necessary by the student's health care provider. (U.S. Department of Education, n.d.) Following the student's time away from studies, they must be returned to the same status they held prior to taking leave. To accomplish this, students may be entitled to make up work, a delayed finish for the semester, and/or automatic readmission.

Employees of educational institutions are also provided leave for pregnancy and related conditions. Employers must, at minimum, provide employees with leave without pay for a "reasonable time" when needed due to pregnancy, childbirth, and related conditions. Following this leave, the employee must be reinstated to the status they had prior to leave, or to a comparable position ("without decrease in rate of compensation or loss of promotional opportunities, or any other right or privilege of employment") (U.S. Department of Education, n.d.) Title IX is enforced via internal Title IX compliance procedures, investigation, and sanction by the U.S. Department of Education, and through private lawsuits.

OTHER LEAVE REQUIRING LAWS

The Pregnant Workers Fairness Act (PWFA), discussed below, requires employers to provide reasonable accommodations, including leave for those affected by pregnancy and related conditions (Pregnant Workers Fairness Act, 2022). The Americans with Disabilities Act (ADA) also requires that employers provide leave, when needed, as a reasonable accommodation for people with disabilities, including pregnancy-related disabilities (Americans with

Disabilities Act, 1990). Finally, many states have their own laws providing benefits to public employees (National Conference of State Legislatures, 2015).

Maternity Accommodations

Pregnant and postpartum people often need accommodations at work or school to protect their health and ensure equitable access to employment or education. Commonly referred to as "reasonable accommodations" or "academic adjustments," they can include changes such as new seating, changes to schedules, lactation breaks, personal protective equipment, and avoiding exposures to teratogens (Center for WorkLife Law, 2023).

PREGNANT WORKERS FAIRNESS ACT

The Pregnant Workers Fairness Act is a federal law that requires employers to provide their employees affected by pregnancy and related conditions with accommodations to how, where, or when their job is done. The PWFA went into effect in June 2023 and covers all public employers and those with at least 15 employees (Pregnant Workers Fairness Act, 2022).

Eligible employees are those who are affected by pregnancy and related conditions, such as pregnancy symptoms and complications; infertility; miscarriage, pregnancy loss, and abortion; childbirth and recovery; postpartum depression; and lactation. To access accommodations under this law, the employee is required to inform their employer of their pregnancy-related limitation. Then, the employer is obligated to offer an interactive process with the employee to determine a reasonable accommodation that would be responsive to the employee's needs. Accommodations are considered reasonable when they do not pose an undue hardship such as added costs or difficulties to the employer in light of their resources.

PWFA prohibits employers from interfering with employees' rights under this law, such as forcing employees to take leave when other options are available, or retaliating against an employee for needing accommodations or asserting their rights under the law. This law is enforced via the Equal Employment Opportunity Commission and private lawsuits.

PUMP ACT

The PUMP for Nursing Mothers Act (PUMP Act), a 2022 amendment to the Fair Labor Standards Act, covers employers of all sizes nationwide

(PUMP for Nursing Mothers Act, 2021). The PUMP Act requires employers to provide their employees with lactation breaks and a lactation space that is not a bathroom and that is free from view and intrusion. Under the law, employees are entitled to take lactation breaks of a reasonable length as often as needed. Employers with fewer than 50 employees total may seek an exemption in limited circumstances. There are no exemptions for larger employers.

The PUMP Act prohibits employers from interfering with employees' rights under the law or retaliating against an employee for needing lactation accommodations or asserting their rights under the law. This law is enforced via the U.S. Department of Labor's Wage and Hour Division and private lawsuits.

TITLE IX

Along with provisions for leave discussed earlier, Title IX also requires educational institutions to provide their students and nonemployee trainees with accommodations/academic adjustments when needed due to pregnancy and related conditions. Federal regulations state that pregnancy and related conditions such as termination of pregnancy and childbirth must be accommodated in the same manner that disabilities are accommodated (U.S. Department of Education, n.d.). The U.S. Department of Education has further clarified that, "to ensure a pregnant student's access to its educational program, when necessary, a school must make adjustments to the regular program that are reasonable and responsive to the student's temporary pregnancy status."

A revised requirement to make adjustments for pregnant students was expected in October 2023 but still remains delayed. The most recently updated draft regulations from the U.S. Department of Education mandate that "[r]easonable modifications to the recipient's policies, practices, or procedures for a student because of pregnancy or related conditions … [m]ust be provided on an individualized and voluntary basis" (Office for Civil Rights, Department of Education, 2022). The new regulations also clarify that students are entitled to lactation accommodations, as lactation is a pregnancy-related condition.

Employees limited by pregnancy and related conditions are also provided a right to accommodations under Title IX. Just as accommodations must be provided to employees with temporary disabilities, they must be provided to employees with pregnancy-related limitations (U.S.

Department of Education, n.d.). Such employees should be provided with reasonable accommodations and are entitled to other benefits or supports available to those with temporary disabilities.

Title IX coordinators are responsible for ensuring that these adjustments/accommodations are effectively implemented. Particularly relevant for STEMM, such adjustments may be required in any educational setting, including field work, laboratory and clinical settings, and externships overseen by the institution.

Anti-Discrimination Protections

State and Local Caregiver Antidiscrimination Laws

Over 200 states, cities, and counties have laws that prohibit discrimination on the basis of caregiver status or family responsibilities for over 50 million employees in the United States (Center for WorkLife Law, 2022). The laws vary in scope, but most prohibit employers from taking adverse employment actions against an employee based on their caregiver status. Some laws cover only parents, or caregivers of immediate family, while others are broader. As of late 2022, six states have these protections for private and public employees; Alaska (protecting parents, Alaska Stat. Ann. §18.80.220); Connecticut (prohibiting inquiries about familial responsibilities, Conn. Gen. Stat. §46A-60(9)); Delaware (protecting family caregivers, as defined by the FMLA, 19 Del. Code §711 (K); Maine (prohibiting familial status discrimination and inquires, 5 M.R.S. §4572); Minnesota (prohibiting discrimination against those living with minors and related inquires, Minn. Stat. §363A.08); and New York (prohibiting discrimination against parents and those living with children, N.Y. Exec. Law §296).

TITLE VII

Title VII of the Civil Rights Act of 1964 (Title VII) is a federal law that prohibits employers with 15 or more employees from discriminating on the basis of sex, race, color, national origin, and religion (Title VII of the Civil Rights Act of 1964, 1964). Under this law, it is illegal for employers to take negative employment actions that are based on an employee's or job applicant's sex, race, color, national origin, or religion.

Discrimination against caregivers is not directly prohibited by Title VII, but caregiver discrimination often comes in the form of sex or race discrimination banned by Title VII. This typically occurs when employers take adverse actions against their employees based on stereotypes or unfounded beliefs about how caregivers of a certain sex or race will act or should act. For example, an employer could violate Title VII by choosing to fire a man who takes time off to care for his ailing parent, based on the belief that men should prioritize work over family care.

Similarly, an employer may violate Title VII by declining to promote or provide opportunities to a mother because of the stereotype that mothers do not want to work long hours or travel. Pregnancy-related bias is also actionable as a form of sex discrimination under Title VII (Pregnancy Discrimination Act of 1978, 1978). This includes failing to accommodate pregnant employees as they would other employees similar in their ability or inability to work (U.S. Supreme Court, 2015).

Discrimination against caregivers can also be based on racial discrimination, which is also illegal under Title VII. Employers may violate the law by treating some caregivers worse based on their race, or making employment decisions based on stereotypes about how a caregiver of a certain race will or ought to behave. For example, it is illegal to allow a White woman to arrive late to work because of childcare issues but discipline a Black woman for doing the same. Cynthia Thomas Calvert found that 8 percent of cases brought against employers for family responsibilities discrimination also alleged racial discrimination, and 2 percent alleged national origin discrimination (Calvert, 2016).

TITLE IX

Title IX prohibits discrimination and harassment on the basis of sex, generally. To that end, it is illegal to treat workers or students experiencing pregnancy less favorably than others, whether doing so is intentionally malicious or not. It is also illegal to base employment or admission decisions on someone's family or marital status. Title IX's prohibition of sex discrimination also requires educational institutions to provide comparable benefits regardless of sex; it would be illegal to provide baby-bonding leave to a student mother but not to a father.

AMERICANS WITH DISABILITIES ACT

The Americans with Disabilities Act prohibits discrimination based on disability for all employers with 15 or more employees (Americans with

Disabilities Act, 1990). Caregivers may be entitled to accommodation and antidiscrimination protection for their own disabilities, particularly pregnant students or employees experiencing complications. Notably, the ADA also prohibits "associational" discrimination, which is discrimination based on an employee's affiliation or relationship with a person with a disability, whether an immediate family member or not. For example, an employer may break this law after rescinding a job offer to an employee on learning their child has a disability, based on the belief they will be less available and committed to the job. Or an employer may violate the law by refusing to allow an employee to take time off to care for a disabled relative, while allowing other employees to take time off for other reasons. The ADA also prohibits harassment based on an employee's association with a person with a disability.

REFERENCES

Americans with Disabilities Act, 42 U.S.C. 12101. (1990). https://www.ada.gov/law-and-regs/ada/

A Better Balance. (2023). Comparative Chart of Paid Family and Medical Leave Laws in the United States. https://www.abetterbalance.org/resources/paid-family-leave-laws-chart/

Calvert, C. (2016). *Caregivers in the Workplace: Family Responsibilities Discrimination Litigation Update 2016*. Center for WorkLife Law, UC Hastings College of the Law. https://doi.org/10.13140/RG.2.2.26004.73608

Center for WorkLife Law, UC Hastings Law. (2022). *Laws Protecting Family Caregivers at Work*. https://worklifelaw.org/wp-content/uploads/2022/11/FRD-Law-Table.pdf

Center for WorkLife Law, UC College of the Law. (2023). *Pregnancy, Childbirth, and Related Medical Conditions: Common Workplace Limitations and Reasonable Accommodations Explained*. San Francisco. https://pregnantatwork.org/wp-content/uploads/Workable-Accommodation-Ideas.pdf

Heshmati, A., H. Honkaniemi, & S. P. Juárez. (2023). The effect of parental leave on parents' mental health: A systematic review. The Lancet 8(1):e57–e75. https://doi.org/10.1016/S2468-2667(22)00311-5

National Conference of State Legislatures. (2015). *State Employment-Related Discrimination Statutes*. https://vawnet.org/material/state-employment-related-discrimination-statutes

Office for Civil Rights, Department of Education. (2022). Notice of Proposed Rulemaking: Nondiscriminaiton on the Basis of Sex in Education Programs or Activites Receiving Federal Financial Assistance. *Federal Register* 87(132)

Pregnancy Discrimination Act of 1978, P.L. No. 95-555. (1978). https://www.eeoc.gov/statutes/pregnancy-discrimination-act-1978

Pregnant Workers Fairness Act, 31 U.S.C. 1115 (2022). https://www.congress.gov/congressional-record/volume-168/issue-198/senate-section/article/S9631-1

PUMP for Nursing Mothers Act. Public Law 117-328 § KK 2022. https://www.congress.gov/117/plaws/publ328/PLAW-117publ328.pdf

Title VII of the Civil Rights Act of 1964, 42 U.S.C. 2000e et seq. (1964). https://www.eeoc.gov/statutes/title-vii-civil-rights-act-1964

U.S. Department of Education. n.d. Title IX of the Education Amendments of 1972. In 20 U.S.C. §1681–1688. https://www.govinfo.gov/content/pkg/USCODE-2022-title20/pdf/USCODE-2022-title20-chap38-sec1681.pdf

U.S. Department of Labor. (1993). The Family and Medical Leave Act of 1993. https://www.dol.gov/agencies/whd/laws-and-regulations/laws/fmla

U.S. Department of Labor. (2023). Fact Sheet #28P: Taking leave from work when you or your family member has a serious health condition under the FMLA. https://www.dol.gov/agencies/whd/fact-sheets/28p-taking-leave-when-you-or-family-has-health-condition

U.S. Supreme Court. (2015). Young v. United Parcel Service, Inc. https://supreme.justia.com/cases/federal/us/575/206/